OUT OF THE WHIRLWIND

OUT OF THE WHIRLWIND

SUPPLY AND DEMAND
AFTER HURRICANE MARIA

Philip J. Palin

ROWMAN & LITTLEFIELD
Lanham • Boulder • New York • London

On the cover: Image of Hurricane Maria by National Oceanic and Atmospheric Administration, GOES-16 satellite.

Published by Rowman & Littlefield
An imprint of The Rowman & Littlefield Publishing Group, Inc.
4501 Forbes Boulevard, Suite 200, Lanham, Maryland 20706
www.rowman.com

6 Tinworth Street, London SE11 5AL, United Kingdom

British Library Cataloguing in Publication Information Available

Library of Congress Cataloging-in-Publication Data
Names: Palin, Philip J., author.
Title: Out of the whirlwind : supply and demand after Hurricane Maria / Philip J. Palin.
Description: Lanham : Rowman & Littlefield, [2019] | Includes bibliographical references and index.
Identifiers: LCCN 2019005087 (print) | LCCN 2019014276 (ebook) |
    ISBN 9781538118214 (ebook) | ISBN 9781538118191 (cloth : alk. paper) |
    ISBN 9781538118207 (pbk. : alk. paper)
Subjects: LCSH: Emergency management. | Crisis management. | Computer networks—Security measures. | Internet—Security measures.
Classification: LCC HV551.2 (ebook) | LCC HV551.2 .P346 2019 (print) |
    DDC 338.5/21—dc23
LC record available at https://lccn.loc.gov/2019005087

To my father, his father,
and Thornton Wilder

Contents

■ ■ ■

Preface

■ ■ ■

But these occasions of human woe had never been quite fit for scientific examination. They had lacked what our good savants were later to call proper control. The accident had been dependent upon human error, for example, or had contained elements of probability.

—Thornton Wilder

REALITY IS BEYOND my ability to fully describe. Yet the effort to discern what is real can be helpful. There are always different angles on reality, each exposing certain aspects, even while obscuring others.

This story is evidence-based but takes poetic license. People and places are mostly composites. There is no town in Puerto Rico (that I have seen) named Montemayor. But what I have described as Montemayor exists at several places in the mountains south and west of San Juan.

Most of my characters are named for characters in Thornton Wilder's *The Bridge of San Luis Rey*. Yet each of my characters also has real lives. I am, for example, one of three living persons embodied (so to speak) in Kathy Juniper.

These poetic devices are my attempt to crystalize what is a complicated story. I have left out much more than I have put in. Applying this triage, I have intended to highlight what I have seen, heard, and decided is essential to reality.

The calendar, weather, inventory counts, and other such details are as accurate as I know how to confirm. The conversations reported are very similar to actual conversations I have had with living principals who I have translated into semifictional characters.

I suspect many readers will be supply chain management students or current industry professionals, all of whom can be especially quantitative. Some have said an algorithm is a very short story. I hope this longer story may inspire new algorithms.

1

Perhaps an Accident

■ ■ ■

Why did this happen . . . ? If there were any plan in the universe at all, if there were any pattern to human life, surely it could be discovered mysteriously latent in those lives so suddenly cut off. Either we live by accident and die by accident, or we live by plan and die by plan.

—Thornton Wilder

AT DAWN ON Wednesday, September 20, 2017, Maria tore into Puerto Rico plunging millions into months of darkness. She swallowed the entire island, chewing it southeast to northwest.

Montemayor is a rural crossroads nearly fifty miles west and two thousand feet above where Maria came ashore. But well before dawn that Wednesday, hot wind brought down trees. Pummeling rain pulled rocks from steep slopes. The Rio de la Plata, already running high, poured over the Comerio dam thick with mud and debris of whole towns. Through this narrow gorge 240 square miles of drainage converges.

Pepita tried to call Jaime. No signal. Her house, even roof, is made of concrete, yet all was shaking hard. She worried over her cousin and his wife in their rooms above the store. The grocery stands high on a ridge, nothing to protect it.

Thirty miles north within the urban sprawl of coastal San Juan, Esteban stood with his coffee behind hurricane-resistant floor to ceiling

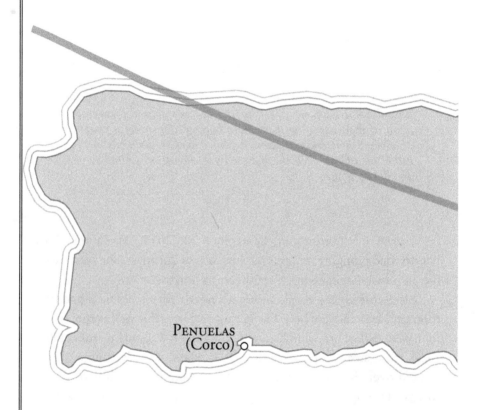

*NORTH
ATLANTIC
OCEAN*

PENUELAS
(Corco)

0	20	40 kilometers	
0	20		40 miles

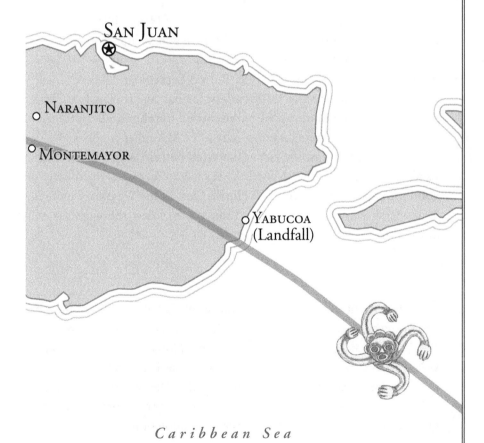

SAN JUAN

NARANJITO

MONTEMAYOR

YABUCOA
(Landfall)

Caribbean Sea

windows. The mountains disappeared into perpetual downbursts. Soon he would lower the steel shutters.

Most days trucks can climb the switchbacks to Montemayor in an hour and a half, delivering Esteban's food, bottled water, beer, and much more. Watching the deluge, he was sure many of the roads to Comerio, Orocorvis, Barranquitas, Montemayor, and beyond were already washed away.

Just before the cell signal dropped, a security guard reported to Esteban the warehouse roof was failing; water had penetrated. Late today or, more likely, tomorrow he would see for himself, not knowing gnawed.

Three hundred miles west, Captain Alvarado eased his barge into the Bahia de Montecristo on the north shore of Hispaniola. Even 400 feet by 105 feet of solid steel needs distance and shelter from a hurricane. Irma's hard hit on Florida—then Jose's erratic course—had kept Alvarado too long at Jacksonville. But on September 13 he claimed a narrow opening for the 1,300-miles-plus passage to San Juan. His impatience had good cause: the barge carried four hundred containers of mostly food, including a jump start of FEMA relief supplies.

Even farther away, the General arrived before dawn at his office. Replacement housing needs for East Texas had increased overnight—good progress on most fronts in Florida. But as the Weather Channel announced Maria's landfall near Yabucoa, he knew the worst of an already brutal season had just started.

* * *

According to the precolonial Taino, our world is balanced between rain and drought, abundance and absence. The volatile goddess Guabancex disrupts this balance if summoned by Caorao blowing into a giant conch shell. The high-pitched whistle becomes louder as it grows faster, accompanied by percussive thumps of distant thunder. The goddess arrives dancing, arms extended in heedless swirl. The character of Caorao is not well understood. But some say he is the personification of fear.

* * *

On September 20, shortly after landfall, the National Hurricane Center reported:

> Maria is moving toward the northwest near 10 mph (17 km/h). A west-northwest to northwest motion is expected to continue through today, followed by a northwestward motion on Thursday. On the forecast track, the eye of Maria will continue to move across Puerto Rico this morning and emerge off the northern coast by this afternoon. The center will then pass just north of the northeast coast of the Dominican Republic tonight and Thursday.
>
> Maximum sustained winds are near 150 mph (240 km/h) with higher gusts. Maria is an extremely dangerous category 4 hurricane on the Saffir-Simpson Hurricane Wind Scale, and it is forecast to retain this intensity while it moves across Puerto Rico. Hurricane-force winds extend outward up to 60 miles (95 km) from the center, and tropical-storm-force winds extend outward up to 150 miles (240 km). A sustained wind of 81 mph (130 km/h) with a gust to 109 mph (175 km/h) was recently reported at Yabucoa Harbor, Puerto Rico. A sustained wind of 63 mph (101 km/h) with a gust to 118 mph (190 km/h) was recently reported at Camp Santiago, Puerto Rico.

Puerto Rican schoolchildren once were taught that their main island is one hundred miles long and thirty-five miles wide. It is a bit larger and there are other islands, too, but the generalization has persisted as satisfactory for most practical purposes. The interior of the island is mountainous with its highest peak reaching 4,390 feet. Since 2008, facing persistent and profound economic difficulties, the population of Puerto Rico has declined by nearly 10 percent. But the island remains one of the most densely populated places on the planet, more so than Britain or Sri Lanka or most of Japan. San Juan's metro region population density is very similar to that of San Francisco or Chicago.

Despite its long-term economic drought, Puerto Rico has the highest gross domestic product (GDP) per capita in Latin America. Puerto Ricans have been early and enthusiastic adopters of wireless communications and all forms of digital commerce. By economic value, Puerto Rico exports more pharmaceuticals and medical goods than the next two mainland states combined. Puerto Rico remains a very rich port.

It is also true that the Port of San Juan receives much more volume than it returns. Over 80 percent of the food consumed in Puerto Rico arrives from Florida's Port of Jacksonville. Sailings typically require six days by barge or three days by fast liner. Of one hundred containers one shipper delivers, seventy-five return to the mainland empty. All the fuel used on the island arrives via the sea. The island was once a huge source of sugarcane, but the Bacardi distillery on San Juan Bay now sources sugarcane from as far away as Fiji.

Since 1973 the government of Puerto Rico has accumulated a debt of $74 billion. This is equivalent to roughly 60 percent of the Commonwealth's annual GDP. (From 1940 to 2017 US national debt has averaged about 61 percent of GDP and in 2018 stands at 105.4 percent.) More than 10 percent of this public debt is related to the Puerto Rico Electric Power Authority, the government-owned electric utility. For a considerable period, and especially since 2008, grid maintenance has been delayed, incomplete, or simply absent. Electric generation, transmission, and distribution were all fragile well before Hurricane Maria.

* * *

In the tropics most mornings start bright. Cumulus clouds often build through the day. But when a hurricane approaches, wispy, icy cirrus steadily accumulate, draping the whole horizon. Birds depart the shore as a hurricane nears. Red ants and seldom-seen subterranean creatures surface. The sea swells and foams. Even a half day before landfall powerful squalls can exceed seventy miles per hour with punishing rain.

Then Caorao calls with his conch. The high-pitched whistle sounds from fluctuations of flow across his shell's edge. Any flow of air creates a whistle when it encounters an impediment. Flow causes the impediment to vibrate. The faster the vibration, the higher the pitch. The greater the flow, the louder the whistle, as if flow can complain of being resisted.

The Taino understood that Guabancex had originally been partnered with her brother Yucajú in creating the world. But when

Yucajú conceived and crafted the first human being, Guabancex was overcome with jealousy. From this sibling rivalry all known evils have emerged. Every human and all that humans create are encountered as impediment by Guabancex. Against these impediments the goddess sends her furious Juracán.

If Caorao is the personification of fear, does his conch call Guabancex or does it warn of what the goddess sends before her? Does the disruption of flow prompt fear or does fear cause the disruption? Perhaps Caorao—fear—does not cause but amplifies fearful consequences.

On September 20, 2017, two flows—one natural, the other human-created—collided on the island of Puerto Rico. The flow of electricity, telecommunications, food, pharmaceuticals, and fuel was dissipated and disrupted by Juracán's destructive dance. Transmission towers collapsed into the tempest. The eyes, ears, and voice of demand-and-supply networks were pulled down with the grid.

Electricity and related tools give knowledge, direction, intention, and strength to contemporary human flows of supply. Absence of electric power is a profound impediment to the resumption of these essential flows.

At higher volumes when flow finds a sufficiently large impediment, the resulting whistle can sound like a scream.

2

Alvarado, Sea Captain

■ ■ ■

There was this strange and noble figure . . . the Captain Alvarado, the traveler. He was blackened and cured by all weathers. He stood in the Square with feet apart as though they were planted on a shifting deck. His eyes were strange, unaccustomed to the shorter range, too used to seizing the appearances of a constellation between a cloud and a cloud, and the outline of a cape in rain.
—Thornton Wilder

IN RHODE ISLAND—and for generations prior—his family had gone to sea for fishing. That is how Paul started with his father and grandfather. But when his dad lost the boat to the bank, he hired on with Boston Towing and Transportation. Thirty-some years later he plies the trade between JAX and various Caribbean ports.

Since 2011 Paul Alvarado has captained the oceangoing tug Cayuco, primarily pulling barges between Jacksonville and San Juan. Barges are not beautiful vessels. They are mostly floating parking lots. Many of the barges pulled by the Cayuco are nearly twice the length of a football field, two-thirds the width, and five stories tall. Containers are rolled on to most barges using trailer chassis and rolled off ready to be hauled by truck. These are called RoRos. Other ships use cranes to lift on and lift off chassis-free containers. These are called LoLos. RoLos or ConRos combine the two methods. Barges are pushed or pulled by tugs. Liners are self-propelled.

The cable with which the Cayuco pulls a barge is made of steel strands thicker than Alvarado's considerable forearm. But even with three solid decks and four hundred containers, not much power is needed to give a buoyant object forward momentum. Stopping is more complicated.

One hundred thirty hours out of JAX, the sun is still below the horizon as Cayuco slips its barge into Bahia de Montecristi behind El Morro point, just beyond Isla Cabra. This anchorage will give both distance and wind resistance in the days ahead.

Alvarado had known the master and crew of the El Faro lost in Hurricane Joaquin. Given the same conditions and a similar vessel, he would have made that run too. Images from their memorial, for bodies never found, were in his mind while planning this roundabout route to San Juan. With at least twenty leagues and a lip of land between him and Maria, the Cayuco, crew, and cargo should be okay and able to make San Juan quickly after.

As the tug and the barge pass tiny Cabra, Alvarado slows the engines, allowing the cable to sag. Then with plenty of slack, he moves the tug about fifteen degrees right from the line of momentum and reverses direction. The cable again draws taunt. At one-fifth power the Cayuco pulls back. The huge barge does not resist and gently slows. As the cable pulls, the tug arcs parallel and then next to the bow of the barge until full stop. Loosening the cable again, Cayuco returns to its forward pull point. Both tug and barge drop anchor.

The sky is clear as the sun now rises behind the tall slopes of El Morro de Montecristi, Dominican Republic, almost to Haiti, four hundred miles from a dramatically different scene in southeast Puerto Rico.

SHAPING AND DIRECTING FLOW

Two-thirds of Alvarado's San Juan cargo is a mix of typical commercial loads: grocery, bottled beverages, various consumer products, and some early pieces of the inventory build for Christmas retail. Many of these containers were originally scheduled for early September delivery to San Juan but were delayed by Hurricanes Irma, Jose, and then Maria. For most of three weeks flow of supplies into Puerto

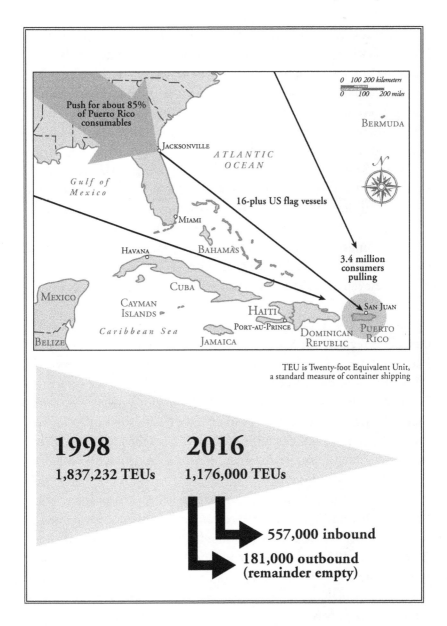

Push for about 85% of Puerto Rico consumables

BERMUDA

JACKSONVILLE

ATLANTIC OCEAN

Gulf of Mexico

16-plus US flag vessels

MIAMI

HAVANA

BAHAMAS

3.4 million consumers pulling

CUBA

MEXICO

CAYMAN ISLANDS

HAITI

SAN JUAN

Caribbean Sea

PORT-AU-PRINCE

DOMINICAN REPUBLIC

PUERTO RICO

BELIZE

JAMAICA

0 100 200 kilometers
0 100 200 miles

TEU is Twenty-foot Equivalent Unit, a standard measure of container shipping

1998
1,837,232 TEUs

2016
1,176,000 TEUs

557,000 inbound

181,000 outbound (remainder empty)

Rico has been impeded, both into and around the island. Two weeks before Maria, Hurricane Irma seriously swiped the San Juan metro region, cutting electric power for more than one million for several days. Sixty thousand are still without power even as Maria arrives.

Most consumer products are transported to Puerto Rico from the Port of Jacksonville by one of three US shipping companies. By law, maritime shipments originating in the United States must be delivered by US ships and crews. In contrast, most of the fuel consumed in Puerto Rico arrives from non-US ports of origin on non-US vessels. The Cayuco—Taino for a flat-bottomed boat—was built in the United States and flies the stars and stripes.

Most San Juan port calls are made by sixteen US flag vessels that regularly cycle between the mainland and San Juan. From JAX to San Juan there are usually about six or seven sailings each week. From both Philly and Houston to San Juan there is one sailing each week. Sailings from JAX require three to six days, Philadelphia takes up to seven days, and Houston to San Juan transit time by barge is eleven days. Then the same vessels return in about the same time.

Prior to 2015 there were at least twenty-two US flag vessels unloading regularly at San Juan. Years before, even more. In 2002 a longtime Puerto Rico carrier sold off its fleet; most was scrapped. In 2008 officials with several carriers were arrested for fixing prices on Puerto Rico maritime transport. In 2011 a major carrier took bankruptcy. Horizon Lines probably should have taken bankruptcy. Instead it was sold in 2015 and the buyer withdrew from the Caribbean. Every year since 2006 the Puerto Rican economy has gotten smaller—lots of churn in these waters.

In the midst of both metaphorical and literal turbulence, what Captain Alvarado can know for certain is that aboard his barge for this September sailing are four hundred containers. Of these, 394 are 53 feet long, 105 inches wide, and 9 feet high. All of these are dry trailers. The other six are twenty-foot refrigerated trailers. Most San Juan sailings include more reefers, but given Maria's threat a last-minute decision was made to minimize fresh cargo.

One hundred thirty-three of the containers are "FEMA freight." The Bill of Lading says simply "disaster relief," but before leaving JAX Alvarado was told the FEMA cargo consists of stacked pallets of bottled water, Meals-Ready-to-Eat, thousands of cots, and a half-dozen industrial-sized electric generators.

The FEMA loads had first been scheduled for early September after Irma hit St. Thomas hard. Most of FEMA's inventory on Puerto Rico had been moved to help survivors in the US Virgin Islands. The Cayuco's containers were intended to replenish the FEMA warehouse on Puerto Rico. Now the contents will be immediately moved to survivors on all the islands.

"What's she doing, Jimmy?" Alvarado asks his Mate.

"Over 150 at landfall, north by northwest at nine miles per hour. Wouldn't want to be in Caguas right now. Closest to us will probably be tomorrow night about sundown, maybe 75 to 80 miles northeast."

"Close enough."

"Plenty of rain, but we'll be port side. Winds should stay below 50 miles per hour other side of the point, not as bad here."

"Double-team lashings on each container. Help me watch the angles. If we can point the bow into the wind, we will. Tell Jose to make the crew's favorites. Minimum watch tonight. Everyone get some sleep."

* * *

Between and behind purple clouds, the morning sun makes its transition from orange fireball to intense yellow disk. Above the Cayuco rises seven hundred feet of limestone, the bright white of its western profile deep in shadow. Crew and cargo will stay behind these rocky walls until Maria is well to the north.

The headland at the entrance to the Bay of San Juan is not nearly so high as at Montecristi. But the mouth of the bay is tighter. Vessels, people, and port facilities are usually well protected. The coast guard and big cruise ships arrive just below the Old City. Barges, liners, and tankers load and unload along large concrete docks deeper into the Bay at Isla Grande and Puerto Nuevo.

Late on September 17, Maria emerged from Saharan heat and Atlantic humidity. On September 18 she destroyed much of Dominica and raced west toward Puerto Rico. Anticipating the inevitable, a Presidential Disaster Declaration for Puerto Rico and the US Virgin Islands was signed on Monday, September 18. FEMA and military assets already deployed to assist in the aftermath of Irma prepared for an even harder hit. All day Monday and most of Tuesday, truck after truck departed the Port of San Juan, moving product in front of the storm. In preparation for Maria, remaining RoRo containers were parked tight side by side. LoLo containers were stacked in dense defensive squares. On Tuesday, the government of Puerto Rico opened more than five hundred schools and other buildings to serve as emergency shelters.

From Tuesday morning, September 19, through eight o'clock Thursday morning, more than fourteen inches of rain fall on San

Juan. In the mountains accumulations are doubled. A wind gust of 109 miles per hour is recorded near the docks. At higher elevation, the San Juan radar dome explodes in winds more than 150 miles per hour. At noon on Wednesday, a two-foot storm surge washes through San Juan Bay.

For eight hours the hurricane crosses Puerto Rico, battering everything north to south, east to west. City streets become rushing streams. Tall steel electrical transmission towers collapse from sustained winds above 140 miles per hour and rain-induced landslides. Between the mountains, Rio de la Plata swells. Flood stage at Comerio is eleven feet. The torrent rises to more than thirty-four feet. As the deluge descends toward the sea every edge is inundated, longtime boundaries obliterated.

Far west, aboard Cayuco the twelve-man crew gathers on the aft deck for an unusual all-hands dinner. The sea is calm. The slopes of El Morro glisten pink and white with the sunset. Wednesday's last rays of sun disappear behind two additional barges anchored earlier in the day.

"Tonight'll be quiet," Alvarado begins. "Tomorrow night could get rough. Friday morning, we should make way for San Juan. Sometime Saturday I hope the Coast Guard will let us in. I'd like to be first."

"How bad is it?" a deckhand asks.

"Plenty. But the cranes are still standing. Our guys at the port are still talking to JAX. It'll be a quick turn. We'll mostly load empties, get back to JAX, unload, reload, and sail back as fast as we can. No candlelit suppers next time. Expect to push extra hard from Friday morning till at least Thanksgiving, maybe Christmas. Get your plates. Eat hearty."

* * *

There is no moon. Beyond the beach, San Fernando's lights are far-off flickers. Soon stars crowd the sky. After dinner Alvarado stands on the bridge deck confirming that Aquarius, Cetus, Pisces, and Eridanus are each in their places.

This is home. He has not been back to Providence since his mother's funeral. His younger brother lives in California. His ex is somewhere near Orlando. His daughter works in Nashville. There's a condo and classic car in Jacksonville, but it's mostly a mailing address. A few years before he and Ana had talked about getting a place together in Puerto Rico, maybe Ocean Park.

At sea Alvarado has often enough confronted death. But it was not until sixty-two, visiting one bedrooms with a beautiful woman of fifty-two, that he began to imagine his own decrepitude. Maybe she did too. In any case, despite finding a nice place overlooking the Parque del Indio, they decided to stick with a room at the Sheraton or her place.

HIDDEN REALITIES

"Looking or listening?"

Alvarado did not even glance. The engineer's tenor was as familiar as sea gulls. "Looking mostly. Strange to stop so long."

"Roger that. . . . What does the sea captain see tonight?"

"How about what I can't see? How about that bright star?" Alvarado points south about twenty degrees above the horizon.

"I see it. What don't you see?"

"That's Fom al-Haut—mouth of the whale in Arabic. Named 1900 years ago. Well, in Greek. Anyway, whaddya think?"

"Bright. Stands out. That part of the sky seems kind of empty."

"Right. I never saw it as a kid. I was too far north. But it was always there. Very consistent star. All the ancient mariners looked for it. I never noticed until Rudy—remember him?—pointed it out one night from a beach bar in Trinidad."

"Rudy always had stars in his eyes."

"Well, even Rudy didn't know there's a planet—Dagon—and probably at least two more—orbiting that star. None of which we see, even looking directly at it."

"Hey, long ways away."

"But right in front of us. What more do we miss from distance or just distraction? You and I are constantly moving between JAX and San Juan. But I haven't noticed Mouth of the Whale since before Rudy died. It's always there. I just haven't looked."

Silence . . . then a faint whistling, sounding far off.

"You see a whale's mouth?"

Alvarado laughed. "Nah. Just hear yours flappin. Everything good below? That piston fixed?"

"Smooth as silk today, smooth as silk. So, I'm going hit my bunk. See you tomorrow. And captain, I don't think you miss much."

* * *

By Thursday morning the hurricane's center was northwest of Puerto Rico, but another four to eight inches of rain with gusty winds swept metro San Juan.

Not much moved that first morning after. The grid was gone along with most phones and public water: no power, no pumping, no pressure. Fallen trees, mud, rocks, and power cables made roads into obstacle courses. Coastal roadways were buried under sand. Comerio, Loiza, Vega Baja, and other riverside towns were swamped under feet of water. Dozens of bridges and even more roads were washed away. Hundreds scrambled to their roofs to escape flooding. Eight persons drowned in Toa Baja. Three elderly sisters in Utuado were killed in a mudslide. Several dams strained, very close to collapse.

Between downpours survivors began to clear roads, patch their roofs, bail and boil water, refuel their generators, and twist moisture from every scrap of clothing. Both temperature and humidity were in the mid-80s. Condensation dripped from concrete, sweat from every pore.

The Port of San Juan had been evacuating since Monday, commercial vessels rushing to get out of Maria's path. On Tuesday night the US Coast Guard (USCG) imposed Port Condition Zulu, essentially closing operations; USCG vessels were sent three hundred miles south. On Wednesday the hurricane snapped navigation buoys, tore up the cruise ship terminal, tossed pleasure craft into San Antonio's pipe, and toppled anything tall and not screwed down very tight. On Thursday afternoon and evening Coasties returned in force to find everything a mess. But after a careful look on Friday the Captain of the Port decided to reopen Saturday for daylight-only operations. At dawn the Cayuco, five other barges, three tankers, and two ConRos were riding the waves within the sight of the famous garita of Castillo San Felipe del Morro.

Each turned for its own place: tankers for the bottom of the bay; three barges to Isla Grande; the Cayuco, other barges, and ConRos for Puerto Nuevo. The flow resuming after three days stopped, after more than three weeks being disrupted.

* * *

"What's the hold-up, Jimmy?" Alvarado scowled. The Cayuco's mate sprinted aboard from the dock beside.

"Not enough trucks. No drivers. Containers are piling-up."

As usual the bright-yellow dock-mules were busy pulling the trailers off the barge. There seemed to be more shunters than usual. But only a handful of trucks could be seen departing the docks.

"What's the deal over there?" Alvarado pointed to five shunters with trailers all stopped in front of a knot of yellow-vested men with clipboards.

"Hacienda is separating relief from commercial. They're expediting FEMA loads for exit, holding the rest for tax payments."

"Shit. Sheep and goats." The captain's left hand pressed to his forehead.

Hacienda is the common name for the Puerto Rico Treasury Department. A sales tax is assessed on the value of specific goods arriving in Puerto Rico and paid before a cargo departs the port. Since 2014 tax payment at the Port of San Juan has been facilitated by a digital system. On Tuesday, September 19 the government, anticipating its digital system would probably fail, distributed procedures for "manual release" of cargo. This required the receiving party appear at the tax office in the port zone and submit a bill of lading, manifesto, bonding documents, evidence of applicable payments, name of consignee, and the cost of merchandise.

For more than four years, none of these had been needed; everything was processed online. The new system was announced just as the grid began to fail. Even if a receiving party had its own truck, driver, and fuel, it could not claim its goods without its own generator, computer, and printer too. On Saturday, September 23, one shipping company advised its customers in Puerto Rico (that could receive e-mail):

> Department of Hacienda's electronic system remains down due to Hurricane Maria. Since approximately 12 pm on Tuesday 9/19 no cargo has been able to be transmitted and no status has been able to be received by the carrier. Cargo that was not transmitted to Department of Hacienda prior to the system closure cannot be processed through normal electronic channels for release. We anticipate based on the devastating impact of Hurricane Maria it will be several days or longer before the electronic processes are back up and running.

The failure of digital transactions—not for the only time—seriously complicated satisfying both demand and supply in midst of the crisis. But even with all the necessary paperwork in hand, during these first days, most of the containers removed from the RoRos, LoLos, or whatever did not move far.

On most days a certain number of trucks and drivers with recurring contracts and security clearances transport containers from the port to specific customers. The delivery locations and routes tend to drift into repeating patterns. Each truck and driver work to find and keep a schedule that satisfies customers and maximizes their financial return. Most truckers in Puerto Rico do not move containers to and from a port. On most days, most truckers are not allowed to drive their rigs into the secure port zone. Especially on an island with an economy that has spent a decade contracting, trucking capacity moves toward a static, even stubborn, equilibrium.

The cargo zones at the Port of San Juan do not usually extract product on weekends. On Saturday, September 23, all Puerto Ricans, including truckers, were still dealing with the recent reality of no grid, no phones, flooding, and many road closures. In the two-million-people plus metro area, every traffic signal was dark. Alvarado cursed the bureaucrats on the dock, but the Cayuco's slow unloading was also the result of less obvious factors. Absence can sometimes elude notice.

By Monday more trucks and truckers arrived at the port, Tuesday a few more. But most of these arrived for FEMA freight. The federal government secured the services of a longtime Puerto Rico logistics leader that was successfully moving relief supplies out of the port to twelve government distribution centers and onto seventy-eight municipalities. Most of the trucks and truckers moving these relief supplies had previously been delivering commercial loads: groceries, meat, produce, construction materials, and other retail products. Setting up the relief supply chain redirected men (mostly) and machines that otherwise would have been available to restore preexisting supply chains.

Further, the relief supply chain is quickly expanding. On Tuesday, September 26, FEMA announced that over the prior six days it had delivered four million meals, six million liters of water, three thousand infant and toddler kits, seventy thousand tarps, and fifteen thousand rolls of roof sheeting. The Puerto Rico relief mission quickly becomes the largest in FEMA's history.

The more vessel space and trucking committed to relief, the more competition and potential constraints on moving groceries and other commercial products. To set sail with 133 containers of FEMA relief supplies, the Cayuco did not load some past-due commercial orders, even as it delivered twice as much commercial volume.

The collapse of the entire electrical grid with cascading loss of related technologies produces uncertainty. What we can usually hear and see clearly is missing. Being suddenly blind and deaf can cause profound perturbations in individuals and systems.

But with retrospective leisure, the situation in Puerto Rico is now clear enough: fundamental demand for food and most other products (other than fuel) was, if anything, slightly reduced by the hurricane. Sources of supply were mostly untouched by the storm. Distribution *capability* was seriously disrupted. Distribution *capacity* immediately after the storm remained essentially the same. But distribution delays—especially during the first week—produced widespread uncertainty that came to be expressed in consumer anxiety and a surge in demand well beyond any capacity of the preexisting distribution system.

One week after Maria's landfall, an estimated nine thousand loaded containers were crowded onto San Juan's docks. Meanwhile, commercial flow out of the port remained but a trickle of ordinary volume. At the end of September many store shelves were empty, home cupboards bare, and delivery trucks were not appearing on their regular routes. Given this absence many observers, not unreasonably, concluded that there were dangerous shortages. Much more was needed, they insisted. The obvious absence was clear evidence that someone (who depended on the angle of observation) was guilty of near-criminal neglect.

Instead, there was an abundance of supply obscured by insufficient—disrupted and distracted—distribution. The absence was obvious, while its cause was concealed in missed signals, good intentions, and unintended consequences.

Even more supply was ordered, and, accordingly, the distribution system was further overburdened and misdirected. As usual, those in most need were most often those left out.

More ships sailed. More trucks were shipped to the island. The port stayed open more hours. Roads were cleared. Fuel was more readily available. Cell phones found their signals again. Slowly, parts of the grid came back.

In August 2017 food stores in Puerto sold $444.9 million in products, just a bit less than in 2016. In September, even with—or perhaps, because of—hurricanes Irma and Maria, food stores sold $460.2 million, much better than in September 2016. October sales grew to $496.8 million. November sales were $512.6 million and December 2017 food sales were $531 million, almost 10 percent higher than in 2016. Some estimate the population of Puerto Rico declined 8 to 10 percent between August and December. If so, fewer people were buying more food and more food was being supplied. Between September 23 and December 23 FEMA also distributed more than forty-eight million free meals.

To deliver this surge, the preexisting fleet of barges and liners was joined by nine new vessels moving between mainland ports and Puerto Rico, increasing load capacity by 40 percent. To assist with the distribution on the island, at least 375 new trucks of various sorts were transported to Puerto Rico in the first two months, many more in subsequent weeks.

Explosive consumer demand in the face of an unusual hazard and uncertain supply is not surprising. It recurs before and after blizzards, hurricanes, and even rumored price increases. It is, perhaps, more surprising how quickly Puerto Rico supply networks—innately narrow and extended—were able to increase both capacity and capabilities to serve surging demand for preexisting channels, even while simultaneously supporting a significant new relief channel. What was assumed to be weak proved surprisingly strong.

But, while this is obvious now, none of this was obvious then or for several weeks. Rather, shortly before noon on Saturday, September 23, with barely one-quarter of Alvarado's containers pulled, unloading came to a sudden and persistent stop. The dockmaster blew his safety whistle and announced there was only room to release and send along the 133 containers marked as Disaster Relief. The rest would need to remain aboard until parking space could be made or found.

For the next few hours Alvarado, his crew, and some unusually enthusiastic stevedores played a frustrating game of Three-Card Monte with twenty-five-ton closely packed cargo containers.

"Aquí está la reina," shouted one of the stevedores, laughing.

"Shit. Sheep and goats," Alvarado cursed again.

3

Jaime and Pepita, Grocers

■ ■ ■

Some say . . . we are like the flies that the boys kill on a summer
day, and some say, on the contrary, that the very sparrows do not
lose a feather that has not been brushed away by the finger of God.
—Thornton Wilder

MONTEMAYOR WAS LONG famous for its fragrant Hoja Prieta
tobacco and smoky sweet Criollo coffee. Tobacco thrives in the sun;
coffee flourishes in shade. Both crops are traditionally harvested
between Christmas and Easter. By Pentecost tobacco is cured, ready
to be rolled and to receive the flame. If kept cool, emerald and scarlet
coffee cherries continue pristine for months after picking.

Jaime, also known as Junior, and Pepita had been raised with
stories of *abuelito* Rodrigo beginning each day with a cigar, cortado,
and banana, all from the rocky soil of his steep four acres and milk
from his own cow. Planted, watered, gathered, dried, roasted, rolled,
and separated by his own hands. Nana called this her husband's
Eucharist: banana and coffee, his mystical body and blood, partaken
in a cloud of aromatic nightshade. The children kept away during
these sacred morning moments.

There had been five children: Jaime's father, also named Jaime,
the eldest; Pepita's mother, the youngest. One infant daughter had
died in the 1919 flu pandemic. Most of the Edenic stories are set in
the 1920s, remembered more from Nana's retelling than any other's

living memory. Jaime, Sr., once confessed, "I have no evidence any of it is true, but I know her stories made me."

The narrative always unfolded with examples of self-sufficient dignity, at least until San Felipe Segundo. Early on the evening of September 13, 1928—the feast day of Saint Philip (aristocratic father and confessor of the gender-bending Saint Eugenia)—a Category 5 hurricane tore through Puerto Rico, killing more than three hundred, continuing to Florida where another 2,500 plus died mostly of drowning. The storm was named San Felipe Segundo because on the very same day in 1876 another deadly, well-remembered storm had struck Puerto Rico.

The first Felipe was bad; the second was worse, washing away Rodrigo's tobacco, taking his barn, and killing his cow. Trees shading the coffee were stripped bare, succulent bushes made brown. No harvest for Three Kings. Only several handfuls of coffee appearing late in Lent. But the banana tree had been saved, wrapped tight with burlap and twine. So, mornings still began with a cigar, banana, and coffee, bitter without milk.

Jaime's father's earliest memories were of hurricane—force wind and rain. But the older Jaime was never sure which images emerged from 1928, when he was five, or 1932 with Hurricane San Ciprian. In 1989 Hugo stayed east of Montemayor. Jaime's father was sympathetic, but unimpressed. Then when Georges hit in 1998, the father wrote the son, "This is what I remember. This is a real hurricane." Jaime is glad his father did not survive to compare and contrast Maria.

Particular events of individual hurricanes were not well remembered. But for the older man, San Ciprian was, by far, the decisive force in his life.

Abuelito Rodrigo's small farm had just about recovered from San Felipe Segundo—he was again taking milk in his coffee—when San Ciprian descended with near-total destruction. In early 1933 Rodrigo and his oldest son left wife, mother, brothers, sisters, and mountains to seed and harvest sugarcane on the coastal plain near Ponce. Remembering the ten-, eleven-, twelve-year-old boy he once had been, the old man considered it the hardest work he had ever done. But there in the south, the illiterate Rodrigo arranged for his son to enter school. In school and on Ponce's busy docks, the teenage Jaime learned English well enough to join the navy at eighteen, meet his California-born wife in postwar Manila, and not return to live in Puerto Rico until 1963 with his wife, navy pension, and twelve-year-old "Junior," who then was tutored in Spanish—and romantic stories—by his Grandpa Rodrigo and Nana.

Three generations and eighty-five years later Jaime, Jr., stands on his grandfather's four acres, considering another hurricane's destruction, and wondering how his life has, perhaps, just been overturned.

Pepita's mother now lives with her, within walking distance of the places and people she has always known. Neither mother nor daughter feels comfortable with English, though Pepita can sing all the words to Beyonce's "All Night" with her two daughters and three granddaughters.

When Pepita's uncle Jaime returned from the navy, he opened a grocery store on his father's property. She worked in the store and grew up with Jaime, Jr. Ten years ago, her cousin came home from a career in the army, took over his father's store, and opened a bakery and café nearby. Pepita is a cashier at the grocery and decorates cakes at the bakery.

Now she watches Jaime and Carlos on top of the second story above the store, lifting and pulling the microwave transmission tower back into place. Dozens of customers are also watching. The doors usually open at eight. But without the tower and its digital signal, Pepita can take only cash for groceries, no credit cards or electronic benefit transactions.

The store had opened the morning after Maria, immediately selling all its ice, even with Jaime enforcing a two-bag-per-customer limit. The shelves of canned goods, limited to three per person, barely survived the first day. But the big Caterpillar generator had turned on exactly as needed. The frozen meat was safe. There was even ice cream, if in the sticky heat anyone could eat it quick enough.

The tower is not wired in place until after 8:15 AM and even then very loosely. The men check that each black cable is firmly joined, hoping they are now reconnected to the preexisting digital world. Briefly considering his desolate horizons, Jaime feels somehow plucked from their green tropical island to a distant dirty brown desert.

Carlos remains on the roof, winching the anchor wires tighter. Jamie descends to open the doors. Pent up heat pours from inside. The air conditioner is off to save power. Pepita edges between the men to her cash register and terminal. She flips on switches. Lights appear and blink. Nothing . . . Nothing . . . Nothing . . . No, just red. They are not reconnected. There must be problems at the other end.

NETWORK DISRUPTION

Montemayor is a rural crossroads. A bit more than a thousand houses are scattered along three intersecting ridges with no specific center. The grocery is near the pharmacy, both of which are a two-minute drive to the bar, bakery, and gas station. The old church is one way; the now-closed school is in the opposite direction. The barrio's population is more than three thousand, but houses are so widely separated that this density is almost hidden.

Montemayor is at the highest peak for several miles along the slope of ancient volcanoes and tectonic uplands that constitute the La Cordillera Central. This range of mountains forms the island's core. Thousands of years of northeastern rains and winds, hurricanes, and earthquakes have carried erosion north to create the coastal plain or spill into ocean depths.

There are three principal roads, each dividing into a spiderweb of other possibilities. From certain places along the Montemayor ridges, San Juan can be seen far below. On another edge there is a steep drop to the Rio de la Plata. Other smaller tributaries cut their own deep chasms. To the south, toward Barranquitas and Aibonito, the landscape gradually becomes drier. But today everything is supersaturated, practically liquid.

"not to San Juan."

"Is there a way to Ponce?"

"Not by Comerio. I haven't heard about Barranquitas."

"Floods to the east. Washouts west. Trees down everywhere."

Most conversations are, one way or another, all about logistics. Montemayor has suddenly become an island within an island, this little grocery its only place of supply. Fridays are typically when the store receives most of its deliveries. But deliveries will not arrive today, not this Friday or for several Fridays ahead.

The lack of signal for the transaction terminal is much more than an inconvenience. Over 60 percent of Jaime's and Pepita's customers use the Family Card to buy most of their groceries. This is the way nutritional assistance is provided in Puerto Rico. Other customers often prefer to use their bank cards to pay, but this is not possible while the terminal's light continues red. The automatic teller machine (ATM) will dispense cash only when it too receives digital confirmations via the roof-top antenna. None of these financial transactions are happening this first post-Maria Friday.

On Tuesday driving his cargo van searching for diesel, Jaime finds a back way into Naranjito. What would usually take thirty minutes takes three times as long. Just as he pulls in, the big cash-and-carry is receiving its first poststorm deliveries from one of the big San Juan wholesalers. He takes all he can fit in the van. Jaime also places an order for Friday pickup: mostly canned meat and bottled water. No ice. No one has ice.

There is a working cell signal in Naranjito that Jaime uses to call his regular suppliers. Again and again the same refrain: "Glad to know you're open. But the roads are awful. But I have lost my drivers. But customer demand is crazy. But I can only do so much, I can't spend three hours driving to Montemayor to feed 1,000, if I can barely deliver to 3,000 one hour away." "Pero . . . pero . . . pero no puedo." Jaime does get promises to put the orders delivered to the Naranjito cash-and-carry on his own wholesale accounts.

While Jaime is foraging for food, water, and diesel, Pepita is at the store. It is Tuesday afternoon. The morning rush is done. Most now know the grocery store is not being replenished. From six short aisles inventory is consolidated into two. Most that remains is now not needed: liquor, frozen food, why did we ever need so many kinds of soap and coffee? A young woman who went to school with Pepita's youngest walks into the store, her two-year-old cradled on her right hip. Pepita knows the family. They buy their groceries and diapers with the Family Card.

Pepita presses the switch on the terminal. The baby cries. Still no signal, the light stays red. The woman looks into the absence where fresh produce was recently sold, her mind not yet fully accepting the reality of mostly empty shelves.

"Carla," Pepita says, "I hope you can help me."

"I hope so too."

"Jaime told me to throw out the vegetables still left. No one wants them. They look so depressing. But in such a time. . . . It seems wrong to simply throw away. Could you take them for me?"

The mother looks at the limp broccoli, pale carrots, and tan lettuce in the refrigerated case, off four days. Entirely depressing.

"Boiled and smashed your Marta won't know the difference. Here, for helping me, please take this stick of butter and this pack of Pampers. Please it would be such a relief to have them gone. What days these have been."

"Yes. Yes. Like the end of the world."

Pepita inhales, one tear escaping as her hand caresses the two-year-old, now fully involved with her mother's earring.

"Or a new opening." Pepita says.

When Nana was not telling stories of the 1920s, she had her own legends of obscure saints and Taino myths. Once upon a time, the erstwhile god, now self-made goddess, Guabancex, sent her hurricanes one after another against her twin brother, high on El Yunque. Again and again swirling spirits lifted the ocean and split mountains, seeking to unseat that sacred sibling, the creator of the human race. He was her opposite: organizing, ordering, defining, naming, making straight, redirecting.

What Yucajú called beautiful, Guabancex found boring, belabored, oppressive, constrained, even tyrannical. He gathered shells, ground them for fertilizer, and used them for roads. She admired each shell's beauty and left them alone. "Tei-toca! Stop!"

the goddess screamed hurling her wild wind and raucous rain against the straight walls and symmetrical gardens of his ivory palace.

Finally, the skies quieted and Guabancex herself, glistening in skin shining like dark gold, arrived at the white mountain to see if her twin had survived. He was not there. His throne sat empty. His pantheon of creative spirits widely dispersed. Guabancex asked their mother, Atabey, if she had seen her son. The Singular Source replied, "I have sent him away for the season. I give you one quarter-turn to unleash and open what Yucajú so carefully tames. May creation learn from each of you and, in the fullness of time, may each of my children learn to love the other."

The young mother's eyes still suggest confusion, but she smiles at Pepita's retelling of the old story. The baby girl coos. Two new customers enter the decimated store.

ADAPTATION

On Wednesday, one week since Maria's landfall, Jaime is returning from Corozal with diesel for the store's generator, three cases of canned milk, and seven cases of white rice, when he sees the bright orange of a Holsum bread truck beside the highway. The driver has stopped for lunch while making his way to a chain store near Naranjito. Jaime pulls his van in front and jumps out. He knows the driver.

"My friend, you have survived, so good to see you."

"Hola, Don Jaime. Is Pepita well?"

"She misses you, my friend. We did not see you last week. No one comes to visit us."

"Yes. Yes. I am surprised to see you. The river road is broken. The mountain road is buried. Is your white van a flying chariot?"

"There is a secret way," Jaime laughs. "Come, I will show you."

"No, No, Don Jaime my delivery is already promised."

"But it is promised to a store on an open road, maybe even a store with a working phone. No? You can find that store again, quickly. But this may be the only chance to bring bread to the poor people of Montemayor."

"No, no. I have my orders."

"I don't ask for the rich man's bread, just some crumbs tossed to the dogs. We are as the beggar at your door, come see our sores."

The breadman shakes his head.

"At least let me show you the way, so that you can come to us tomorrow."

The breadman purses his lips.

"It is not far. It will not take long. I know your boss, I will write him a letter reminding him that little Montemayor also needs its daily bread."

The trip takes longer than Jaime anticipated. There is a half-fallen tree too low for the bread truck. It takes more than an hour to clear. But the breadman leaves four racks of loaves and returns Thursday with sixteen more. On Friday there is a whole truck, all sold in less than three hours.

By Friday, September 29, the grocery has four rows of mostly stocked shelves. The regular wholesalers are now delivering to the Naranjito cash-and-carry to which Jaime travels every other day. Bread deliveries have restarted. Pepsi and Coke have each delivered once. Sales for the week after Maria, Friday to Friday, are two times normal. Pepita calls it buying-down-worry. The customer needs only two cans for the week but buys six, just in case, and then comes back again the next day.

Two weeks before Maria—restocking after Irma—Jaime had filled his freezer with nearly $50,000 worth of meat. Carlos now wears four layers of cotton clothes and a heavy black garbage bag to cut slabs into meal-sized portions, every half-hour or so emerging from the freezer in a white fog, delivering the rock-hard meat to waiting customers, warming his hands before disappearing into the cold again.

Propane has become the eternal flame. Every home with a gas grill is now a place of communal worship. Barbeque, derived from the Taino *barbacoa*, once again gathers the entire tribe.

If not for the cost of fueling the generator and long hours driving back and forth for supplies, Jaime could have considered

Maria a commercial benefit. In typical times many of those living in Montemayor shop at Jaime's store only as a convenience. Those who work off-the-ridges buy most of their groceries at one of the chains near their places of employment. There they find more variety and often slightly lower prices. The majority of customers Pepita serves day in and day out are old or poor or both, people who do not have transportation or money for gas or confidence to drive narrow, twisting roads to elsewhere. But now even rich women are standing in line waiting for Carlos to appear out of his freezer like a genie from a bottle.

By his second week without the grid and related disruptions, Jaime has settled on what the retired US Army sergeant major considers his battle rhythm. On Tuesdays, Thursdays, and Saturdays, he will take the old Ford van fitted with the 250-gallon poly tank to pick up diesel from his buddy in Palomas. The generator sucks about one hundred gallons per day, so this will keep at least one day and a bit more extra fuel available. On Mondays, Wednesdays, and Fridays, he will drive the International 4300 to Naranjito to pick up wholesale deliveries. On Sunday, he will change the generator's oil and reconcile financial accounts. Jaime gives constant thanks for Pepita's fastidious tracking of credits. He is not nearly as rigorous or confident regarding debits. What he orders, what is invoiced, and what is received can be dramatically different.

Without the transaction terminal, commerce depends on scads of cash and scraps of credit. Jaime trusts Pepita's judgment on who needs help and who they can afford to help. So far, she is keeping sixteen pages—one page for each customer—of accumulating debt to be paid in thirty days. The first entry had been opened for the elderly widow living next to the store and the most recent for a rich man who has run out of cash. After one week, the running total for each of the pages ranges from $22.23 to $140.63. Jaime recognizes each name and understands why Pepita has visited Don Fiao on each of these neighbors. Jaime doubts there is a bad page in the set once EBTs, credit cards, and ATMs return. But when that might be is not yet obvious.

Jaime is also handing over wads of currency. While the wholesalers deliver to Naranjito on credit, he still has to pay cash for diesel—roughly $300 to $400 per day—payroll, gas for the truck and van, and cash at premium prices for hard-to-find products such as diapers and ice. Ice could claim a king's ransom. He also has to keep back enough cash to eventually pay his wholesalers, assuming he can resolve what they say he owes and what he is sure has been received.

On the long drives for diesel, food, bottled water, and what-not, it occurs to Jaime that most of his current abilities emerged from prior vulnerabilities. The big poly tank welded to the van's back end had originally been purchased to transport water during the 2015 drought. The beautiful International 4300 insulated box truck with Thermo King refrigerator had been signed over to him in August. His younger brother could no longer make the monthly payments. Jaime took over the loan balance. It was, he thought then, much bigger than he would ever need. Now every square foot is being filled three times every week. If the electric utility had been more reliable, Jaime would never have suffered the expense of buying and maintaining the preowned Caterpillar generator. Prior problems prepared him for the big problem of Maria and even bigger problem of weeks without the power grid.

By the end of September, eleven days after landfall, Jaime is sore. Pepita wakes up in the middle of the night with horrible nightmares. Water is scarce. But people are being fed. Many customers are keeping medicines in the store's one-time dairy cooler. Neighbors who had barely talked to each other are sharing grilled meat and watching the sundown together.

Many roofs and whole houses have been lost. Below Montemayor, along the Rio de la Plata, flooding was catastrophic. The constant heat is exhausting. Those already ill are now weaker. Many who had been fine, now are not. But even by the end of September, Jaime begins to perceive that, perhaps, for him—as for his father and grandfather—hurricanes both open and close.

On his fiftieth birthday, Jaime's wife, Ruthie, had given him a hand-carved plaque with a quote from *The Bridge of San Luis Rey*: "Jaime tried to conceal his shame for he knew that one of those moments was coming that separated him from other people." Ruthie found the quote both accurate and hilarious. The future, good and bad, often provided him previews, not in any detail but with a strong sense of direction and outcome. This inner vision, perhaps voice, had well served the private, then sergeant, then sergeant major in Vietnam, in Iraq, and again in Afghanistan. Ashamed or not, it visited him now where his people had called home for four hundred years— Nana claimed longer. All were tough times; Puerto Rico post-Maria is, for Jaime, the toughest place yet.

* * *

But on the first Monday in October Jaime feels like a hero. Twenty small grocers have combined resources and sent one of their members to Santo Domingo. There he secures ten 40-foot reefers full of ice and vessel space on a small inter-island barge. At dawn on October 2, Jaime is on the docks at Ponce in the Thermo King to claim his portion. Shortly before noon the truck is straining up the final steep mile to Montemayor when a line of cars ahead causes him to stop. Another landslide is Jaime's first thought. But then the passengers in the two cars in front of him jump out, obviously celebrating.

"The iceman cometh!" a teenage boy with goth-cut hair announces.

"Some people are worth melting for!" his mother proclaims.

"Él monta al rescate!" a driver shouts.

The entire barrio has heard that Jaime left in the dead of night to get ice. Now half those with cars have converged on the three roads that meet at his store, blocking traffic. Several cars pull into the left lane in front of him, horns honking, headlights flashing, hands waving out the windows, and leading the white box truck up the hill. About one hundred yards from the store, smiling members of the local constabulary direct Jaime and his escorts into the park and onto the ball-field. Crowds come laughing, running, and carrying their ice chests. Pepita appears with the cash box. In less than two hours nothing is left but warm puddles.

PERTURBATION

On Monday night Jaime sleeps better than he has since Irma. On Tuesday morning, he even takes time for a cortado. About 10 AM, he starts for Palomas, five miles away, to pick up another 250 gallons of diesel. A high school buddy operates his own independent gas station and mini-mart. On Monday morning Jaime had sold him a half-pallet of ice on his way back to Montemayor (and filled up his truck with diesel). His friend's station was one of the few in the area with an emergency generator and has been open since the day after landfall. Twice in the past two weeks, the station could not sell Jaime a full 250 gallons, but each time he was back in two days and made up the difference. The station is supplied from racks near Tallaboa, which, Jaime guesses, are not being drained as fast as the fuel terminals near San Juan. There is a chain gas station closer to the store served from San Juan that has been closed as much as open.

At the Palomas mini-mart, there are orange cones in the vehicle bays. Signs are taped to the pumps: "No Hay Gasolino."

His friend walks out to the van, "Bad news. Nothing yesterday."

"Today?"

"Don't know."

Jaime quickly calculates that his store's generator and reserve tank have two days of diesel, maybe a bit more. He drives on to Aibonito. By three o'clock he is next to the tollway in Cayey, refilling his van's gas tank. But no one has diesel they are willing to sell. Some truckers at the Burger King say diesel is dry all the way to Guaynabo. One of the drivers complains FEMA has taken over the ports and is not letting out fuel or food unless mayors request it. An EMT says he saw diesel deliveries made to two hospitals this morning. Maybe tanker trucks have been commandeered.

Jaime does not find any diesel on Tuesday. Before renewing the hunt on Wednesday morning, he tells Carlos to stop selling meat at 11:00. Jaime tapes up a sign saying the grocery will be closed for the afternoon and will reopen for regular hours on Thursday. Pepita can work with Ruthie at the bakery. He tells Carlos to turn off the generator at noon, but watch the temperature gage, turning the generator back on whenever the freezer's temperature exceeds 36 degrees. Jaime hopes to be back before the generator is restarted. He hopes his amigo in Palomas has been resupplied. But, just in case, this on-and-off should extend the fuel, even if it stresses the engine.

As he opens the van door, Jaime is still not sure if he will head north toward San Juan or south toward Ponce. He is almost sure the most diesel arrives at the Puma and Total terminals near Port of San Juan. But the metro region is also where the greatest demand exists. If there is lack of supply, San Juan would, he decides, probably feel it worst. There are two smaller fuel terminals west of Ponce, including the rack that had regularly been the source for his friend in Palomas. There is even a rumor the grid has been restored around Guayanilla. Jaime turns south to check at Palomas. No resupply has arrived. He continues southwest toward Barranquitas, then Villalba, and then Ponce, planning to drive until he finds some diesel.

In many places the highway narrows to one lane past a landslide or missing roadway. Long lines extend in both directions. Detours meander with no arrows, vehicles depending mostly on the flow of other vehicles and what the sun's position suggests might be a reasonable choice.

There is no diesel in Barranquitas, though Jaime does see the tall tower with which his store's aerial had usually been in dialogue. It is still prostrate in the street. In Villalba there are diesel tankers under guard. The fuel is reserved for utility crews and mainlanders working the transmission lines along the northwest edge of the city. On the outskirts of Ponce, Jaime asks about available diesel. He is sent in many different directions, none of which uncover any for sale. As the sun disappears, he turns off the main road toward Penuelas to spend the night, close to two nearby fuel terminals.

Jaime smells his own stench mixed with that of the filmy residue of fuel sticking to the polyurethane tank's interior. There is a hint of mildew, probably from the seat fabric. He retrieves a sleeping pad and soiled pillow from behind the passenger seat. Before stretching out on the sandy soil, Jaime buries an envelope beside a tree. It holds $4,000 in small bills. He keeps $400 in his wallet as a credible diversion. Reclining, Jamie recalls his father reading aloud and laughing as Don Quixote holds his nose at Sancho Panza's unsavory smell of fear, to which the everyman replies, "Why do you lead me a wild-goose-chase, and bring me at such unseasonable hours to such dangerous places?" Whether or not the threat's as bad as Iraq, he smells even worse. Soon it is inky dark. Jaime is exhausted but does not sleep soundly. It has been two weeks since Maria's landfall.

Grinding truck gears wake him at dawn. The rusty remains of the CORCO oil refinery fill the hillside with tall steel smoke stacks and intricate piping, not unlike the giant columns of Saturn's Temple rising beside the Roman forum. Each ruin offers clues to the strength and weakness of their builders.

When Jaime was in high school, 80 percent of the fuel consumed in Puerto Rico was produced here. CORCO was the second-largest employer on the island and a US Fortune 500 company. But change in global energy markets after 1973 was too fast and fundamental. The company failed and closed in 1982.

Cannibalizing surviving infrastructure, smaller players now share the site: two fuel distributors, the Costa Sur Power Station, an LNG importer, and tankers still anchor at the jetty extending into the bay, but they deliver refined, rather than raw, petroleum. Outside San Juan, there is no place better to find diesel on the island. Jaime pulls his van outside a gate across from derelict crackers and cokers. Tanker trucks are already lined up.

But while there is diesel, Jaime is not allowed to buy it. One guard says they are not selling without a preexisting contract. At another gate a dispenser skeptically examines the van's Mad Max apparatus. He says stateside Jaime would be arrested on suspicion of terrorism. To fill him would cost the dispenser his job. Jamie talks to truckers, but none in the line are delivering to retail. He returns to Montemayor by a different route, checking all along the way, finally again at Palomas. All are empty. He should have gone north.

"Any idea why?" Jaime asks. "Why not the first week, but now?"

His friend shrugs, "More stations are open. Well over half. In San Juan even more. More mouths on the same tits. Runts get pushed away."

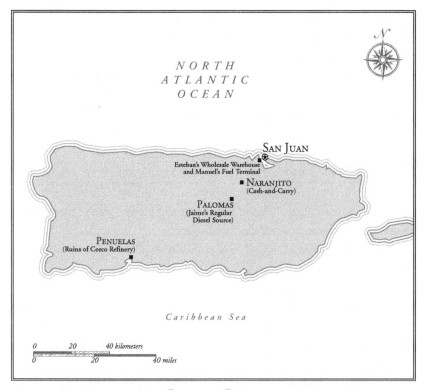

PUERTO RICO

Very early Friday morning the generator gives off a high-pitched sigh as it burns the last of its diesel. On Sunday at the same park where he had sold ice, Jamie gives away nearly $30,000 worth meat at a community barbeque. The Baptists collect a free-will offering of $3,000, but Jaime demurs.

FLOWS FUNNELED

As you travel to Montemayor on the road ascending from Rio de la Plata, a twelve-inch-diameter steel water pipe runs parallel. At the edge of several switchbacks are pumping stations, often with an attached emergency generation unit, and large pressure tanks. The system pumps water more than 2,200 feet in elevation.

This ambitious rural public water network is the result of human need, population density, and a stubborn New Dealism persisting well into the 1960s. Over 95 percent of the island's population is served by the Puerto Rico Aqueduct and Sewer Authority. But while administratively and financially centralized, functionally the aqueduct is a loose network of fractured subsystems. In the case of Montemayor, local households consume roughly forty-five gallons of water per day drawn from a filtration and pumping station near the Rio de la Plata. Well before hurricanes Irma and Maria, there was a decline in the water system's reliability and safety. Public water is widely used for flushing, washing, and bathing, but health concerns prompt most people to purchase bottled drinking water.

During and after Maria the riverside water plant was overwhelmed by more than thirty feet of water pulling every lost car, fallen tree, ounce of mud, grain of sand, parasitical strain, and toxic compound from across the whole watershed. $P = F/A$, pressure equals force divided by area. Force equals mass multiplied by acceleration. Thirty-plus inches of rain falling over 240 square miles is a huge mass. For this mass to descend over high mountains into narrow valleys in less than forty-eight hours produces extreme force. Intake valves were broken, pumping equipment ruined, and pipelines fouled. Water is essential to human life. Too much in the wrong place at the same time can be deadly.

On the first Saturday in October, National Guard trucks and equipment arrive. It is a combined detachment from Puerto Rico, South Dakota, and Ohio. They deploy large reverse osmosis water treatment devices. In combination they can produce about 4,500 gallons per hour of drinking water, roughly 90,000 to 100,000 gallons per day. The South Dakota unit has big beige bladders to hold supply. Local government water trucks distribute to residents across the municipality. Before the storm just Montemayor's households— about 10 percent of the municipal total—were using at least 60,000 gallons of water per day. But still, this water mission is a critical contribution to human hydration.

Reverse osmosis uses filters and specifically applied pressure to undo the general dispersal of solutes in a solvent, usually to remove unnecessary or unwanted elements from the principal flow. The filter or filters keep back what's not wanted, allowing through what's most needed. In the big National Guard reverse osmosis engines, the energy of the purified water flowing out is captured by a turbine (or sometimes a piston) and contributes most of the energy needed to keep applying pressure.

The National Guard brings bulk water to Jaime's Sunday barbeque. They leave behind a mid-sized bladder to serve as a regular point of water distribution. On Monday three men from the municipal mayor's office accompanied by two police officers appear in Montemayor. They arrive with a truck-load of FEMA snacks, hundreds of meals-ready-to-eat, and park beside the National Guard's water-bladder. Soon a crowd arrives to receive the food. Jaime notices the crowd lining up as he returns from Naranjito with his box truck full of canned meat, canned spaghetti, canned peaches and pineapple, canned beans, canned soup, and more cases of canned green beans than he really wanted. He also has twenty cases of bottled water. As Jaime and Carlos unload, Pepita is busy at her check-out, penciling into a spiral-ring binder each transaction. Despite no frozen meat, the number of customers seems about the same as the week before.

On Tuesday morning, October tenth, diesel returns to Palomas. Jaime's generator growls back to life. The grocery aisles have lights again. With nothing to freeze, the cooler is set at 74 degrees to hold off mold and to chill water and beer. Pepita checks the transaction terminal, still no signal. Early afternoon the mayor's men return with more FEMA food and now with bottled water. With less pull for cooling, the diesel lasts longer giving Jaime more time to focus on food. He even finds some frozen pork to sell. The third week's battle rhythm adapts to changing conditions.

In the decade since Jaime took over the store from his father there has been a rough balance between push and pull. During the first few years, as the flow of customers slowly increased, he noticed their habits, adjusting what he bought—and how much and when he bought—to what he found would sell. The store expanded its inventory of beer, wine, and liquor, while cutting back on fresh vegetables. Bread and meat were mostly sold on weekends. Jaime hired Carlos to cut meat when he realized the store could compete with the chains on price and beat them on quality, if he carefully timed his volume buying. Since Maria, though, none of these lessons apply. Consumers are so uncertain that they buy almost anything available. Despite the free FEMA food, the store's revenue for the second week in October is double the second week in September and three times the second week in August.

What worries Jaime most is some sudden shortage in the flow of cash. Before Maria less than 30 percent of store purchases were made with paper money. Since September 20, except for the small role of Don Fiao, all purchases have been cash-only. At the start of October, after paying expenses, Jaime had accumulated more than $23,000 in rolls and rolls of small bills. Part of his motivation for cashing checks has been to reduce the risk of so much currency on hand. But especially since diesel suddenly disappeared and then mysteriously reappeared, Jaime wonders how long this financial flow can persist. How much is he risking when exchanging cold cash for a potentially hot check—or even a good check—if there is no bank open for processing? After thirty days will he need to appear at his wholesalers with bags full of one- and five-dollar bills and 2,000 quarters? When will too many people no longer have any cash? How long can a cash economy continue on fumes? Jaime remembers the mournful whistle his generator emitted sucking its last dregs of diesel.

* * *

On Friday, October 13, only four cars and six people are waiting for the doors to open as Pepita arrives shortly before eight o'clock. Jaime is busy preparing to take the truck to Naranjito. Carlos is walking toward the store just past the baseball field. Ruthie unlocks the doors from the inside.

Lights are already bright. Four of six aisles are well stocked. When Jaime returns the few gaps will be faced front. Mrs. Ortiz, the widow next door, places one can of condensed milk and two cans of Carmela chicken sausages on the check-out. As Pepita reaches for her spiral notebook, she presses the switch on the transaction terminal, as she has done every day. Mrs. Ortiz mumbles some dark humor regarding the day's bad luck. Pepita does not get the joke, but still smiles and nods. There's an odd beep. Pepita glances toward the transaction terminal: green.

Pepita asks Mrs. Ortiz if she has her Family Card. The old woman's eyes swell and redden. She shakes her head, thinking Don Fiao is dead and she cannot make her purchase. Two others are in line. Pepita asks, "Do you have a card? Any card?" Carlos enters as Pepita swipes a bankcard from Mrs. Suarez. Another beep. The green light blinks. Confirmed. Pepita gasps.

"Sistema! Sistema! Sistema!" she shouts. "Sistema! Sistema! Sistema!" she shouts again and twirls behind her register. Carlos bounds to the door and calls Jaime to come quick.

4

Esteban and Manuel,
Distributors

■ ■ ■

What relationship is it in which few words are exchanged . . . in
which the two persons have a curious reluctance even to glance
at one another; and in which there is a tacit arrangement not
to appear together in the city and to go on the same errand by
different streets? And yet side by side with this there existed a need
of one another so terrible that it produced miracles as naturally as
the charged air of a sultry day produces lightning.
                                                              —Thornton Wilder

IN 2016 PUERTO RICO consumed an average of 42,000 barrels of gasoline per day. About 45 percent of fuel is off-loaded at San Juan Bay,
30 percent at Guayanilla Bay, and 25 percent at Yabucoa Bay. When
Maria made landfall at least thirty days' gasoline supply was on the island
(several sources suggest up to sixty days, depending on how consumer
demand is calculated). In the first nine days after landfall eleven fuel
tankers unloaded 2.4 million barrels of gasoline, diesel, and other fuel.

Before the grid collapsed, about half the electricity consumed
in Puerto Rico was generated by burning diesel or residual fuels.
Without a transmission network, the electric utility no longer consumed this flow. Preexisting diesel supply and distribution channels
became available to feed the new demands of emergency generators.
But distributing diesel to thousands of scattered residential, institu-
tional, and commercial customers is an entirely different time-and-
space equation than delivery to a handful of power stations.

Most smaller generators use gasoline instead of diesel. In the ninety days after Maria's landfall, retail gasoline sales in Puerto Rico increased by about 12 percent. In January 2018 gasoline sales peaked at 32 percent higher than the previous January. In February 2018 gasoline sales began to decline. By April 2018 retail results had returned to prestorm patterns.

In 2017 the electric utility had 1.5 million customers. Since 2013 electricity customers in Puerto Rico have, on average, lost power five times per year for about three hours each outage. Over the years customers have accumulated a diverse set of larger diesel, smaller gasoline, and propane backup generators to manage these occasional electrical outages. Most of the gasoline units are self-serviced. Backup systems have been fueled by a small fleet of mostly independent site-delivery services.

On September 19 there are roughly 60,000 customers who have been off the grid for almost two weeks. The next day the number-not-connected is multiplied by twenty, encompassing every electric utility customer on the island. Not until the end of November are more than half of customers back on the grid. Not until early January does the proportion exceed 70 percent.

For smaller home generators it was possible—problematic, but possible—for tens of thousands to appear at retail gasoline stations to be supplied. Before September 20 there were a few more than 1,100 gasoline retail stations in Puerto Rico. By September 27 roughly five hundred had opened. After two weeks at least 750 were operating. One month later about 850 were selling fuel.

DISEQUILIBRIUM

Especially in the first two weeks, being open did not mean having gasoline for the full day. A station could receive a typical week's worth of gasoline and be empty in a few hours. To give those standing in long lines hope of receiving something, many retailers rationed purchases to no more than fifteen dollars, not even enough to fill larger jerricans. This may have been a wise social decision. But it guaranteed seemingly infinite lines. An efficient 2,000-watt portable generator—enough for a refrigerator, small appliance and lights—will burn about a gallon of gasoline every four to eight hours, depending on draw. A 5,000-watt generator will burn about one gallon per hour. With a ten- or fifteen-dollar purchase, most customers had much less than needed for a single day.

On September 27, 1.5 million households were off the grid. Fortunately for the fuel distribution network most of those without grid connections did not have backup power. No one really knows how many residential electric generators were on the island that first week. The head of a Puerto Rican association of electricians conservatively estimates there were at least 100,000.

Gasoline is sold by liter in Puerto Rico, but for metric-challenged mainlanders, let's peg gasoline at three dollars per gallon with a maximum purchase of $15.00, so five gallons per customer per trip. There are many different generators but average consumption of ten gallons per day is reasonable. Given rationing, this means operating each generator requires at least two buying trips per day. Because people are understandably nervous, three trips are likely. There may be an average of five pumps per station.

Two hundred generators per station, with rationing, translates into six hundred buying trips per station per day and 120 purchases per pump per day. This adds up to a real demand of six hundred gallons per pump per day or three thousand gallons per station per day (before refilling any trucks or cars). While storage capacity at gasoline stations differ, stations are refilled by tankers that typically carry nine to ten thousand gallons.

Prior to Maria, most gas stations were resupplied about once each week and seldom required a whole tanker load. There were enough fuel tankers on the island to deliver supply for this recurring demand. Post-Maria the real demand for generator fuel—even before refilling vehicles—is more than double preexisting retail distribution capabilities.

To add a further dimension of time: Assume six minutes for each customer to physically fill and pay for their five gallons. In sixty minutes, each pump can deliver fifty gallons. On September 27 there

September 27 Fuel Flows

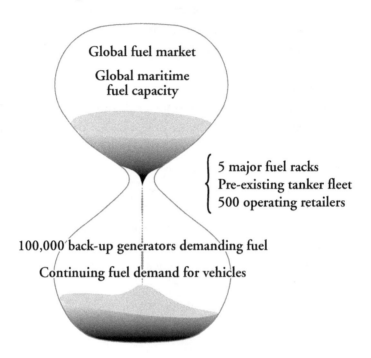

Global fuel market

Global maritime
fuel capacity

$\left\{ \begin{array}{l} \text{5 major fuel racks} \\ \text{Pre-existing tanker fleet} \\ \text{500 operating retailers} \end{array} \right.$

100,000 back-up generators demanding fuel

Continuing fuel demand for vehicles

is a curfew from 6 PM to 6 AM. Over a twelve-hour shift six hundred gallons can be delivered by a single pump. This happens to match projected demand for backup generators. But any increase in average time to make a purchase—for example, to allow a truck to fill up with gasoline—will leave customers with real need without fuel. No wonder the lines seem unending.

A FUEL TERMINAL

On September 20, sun and moon are in conjunction, not even a thin crescent showing. Thursday's predawn seems doubly dark as Manuel makes the white knuckled drive from his home to the terminal, making several detours before finding a way through the maze of destruction. But the expressway is passable with almost no traffic. The intersection to the tollway is flooded and the next exit too. Fortunately, Manuel is driving his wife's SUV, not their Mercedes.

At the terminal west of the port, standing water is everywhere, ankle deep or higher in many places. The office roof has partially peeled. The executive offices on five are floating, accounts receivable on four is soaked, and marketing on three is dripping. But the hurricane windows held. As the day-after dawn starts, Manuel assembles his team on the second floor.

"No structural damage to the tanks that we can see. Domes are secure."

"Normal pipe pressure."

"One of the generators is running rough. But all are online right now."

"No cell signal, no land lines, no internet. Our servers are fine, but customers can't access the purchasing system. No EDT connections."

Esteban exhales with an involuntary low whistle.

"There are three tankers at the gate. Security says they have paper purchase orders, but there's no way to digitally confirm or take payment."

"What about the dock?" Manuel asks. "Can we still off-load?"

"No word yet. I sent Maxie to check and report back."

"Other than us there's the five-man night-crew," the terminal manager reports. "Victor Garcia is still checking the racks, Juan Nieves arrived just as I came upstairs, I sent him to check the propane compound. Brenda is in the lab checking water samples, so far no trace of product leaking."

The terminal manager's radio crackles. It is 7:30 on the morning after landfall. Seven tankers are now at the gate. Fewer than usual for a Thursday, but ready to be filled. There is, however, no immediate means available for them to pay. Supply is in place. Fuel infrastructure is intact. But the high-efficiency, high-volume, digital transaction system is missing in action.

One of the truckers has told the guards that their principal competitor's terminal, less than five miles away, is not expecting to open today.

"What about CORCO? Peerless?" Manuel asks his team.

Heads shake, shoulders shrug. No one knows.

"I can't imagine Buckeye is open," Manuel mutters, mostly to himself. There are five major fuel terminals in Puerto Rico. Buckeye is at Yabucoa, immediately adjacent landfall.

"We may be all that's left," Manuel says looking out on the seventeen huge tanks. More rain is falling through weak light of a shrouded sun. The usually grassy field is a watery expanse. Safety lagoons around each tank are filled to their brims. White towers rising above the flood, filled with over eighty million gallons of fuel.

Manuel turns to the CFO, "We need a paper process to get the tankers moving through the racks. I can't imagine we'll see many more today. But tomorrow . . . We have to be ready. I should try to call London."

Manuel has been in the energy industry his entire life. Before returning to his native Puerto Rico as general manager, he had assignments with two different multinationals in Mexico, Honduras, and Chile. He has dealt with refinery explosions, pipeline spills, and evacuating drilling platforms. On the eve of landfall, Manuel warned his wife this could be different.

"Maria will have an all-island effect. The whole network could cascade. The grid almost certainly, then telecoms, the water system. If damage to the port and roads is too bad, I don't know how we come back. The terminal could be fine, but without the port there is no flow. With no flow there is stasis. Stasis is death."

Manuel had briefly considered spending the night at the terminal. Instead he stayed home. After dinner they called their daughters, one in Miami, the other in Madrid. He did not sleep soundly.

From the roof Manuel dials the UK cellphone number for the Managing Director Americas.

"Hola, Manuel. It is good to hear from you," Duncan's Scots lilt sings even through this tinny satellite transmission.

"Terminal is good, great given context," Manuel begins. "Not yet certain about the dock, but pipeline pressure is reassuring.

"At dawn's first light Landsat gave us a passing peek at the bay and La Curva. Nothing obviously bad. Maritime chatter is tentative, but nothing discouraging. We have usual contacts with Coast Guard. Call back in an hour, I may have more."

"Hope I do too. One big impediment: No digital transactions."

"Aha, of course. . . . That's sticky."

"Very and our competitors do not seem to be bouncing back. So, I expect tomorrow we will have several first-time customers."

"What's your thinking?"

"First, we fill our own tankers and get them to every one of our retail stations that can pump. Branded retail has priority. Second, Antonio is jiggering our plans for non-digital transactions. I want to fill every tanker that arrives, prior customer or not. I want to maximize flow. Maybe we give prior customers their own lanes, but we serve everyone."

"What are your principal risks?"

"Near-term is resupply. Longer-term is cash-flow. Without resupply, I am accelerating toward full-stop. Then, instead of payments made before each tanker leaves the terminal, we are going to have a mess of paper, months of invoicing, contentions with long-time customers, deadbeats who never pay, and lots of legal fees."

Duncan laughs, "So, your biggest worry is me. Will I get you resupply, and will I punish you for a job well-done? There's plenty of resupply already in the water waiting to enter port. Did the jetty survive? Can we restart flow? We don't know yet, but I bet yes. Then regarding cash-flow—may that fierce god bless us all—real concern. But a rare *benefit* of being but one bluish vein in the huge hairy beast of our employer, is the super-abundance of cash-flow. Whatever your loss-of-flow these next few months, the beast will still rampage. More importantly—more seriously—you and I both know two things: First, this is a fantastic opportunity to claim market share. Second, it is the right thing to do. Not often in our lives are we given the chance to do so well, by doing good. Glad you're there to make it work. I'll do my job. You do yours."

That brutal day-after only sixteen tanker trucks arrive at the gate. Manuel's team fills and releases all of them. But without digital transactions, the typical service rate of fifty-five minutes gate-in to gate-out is nearly doubled, all the additional time required to record identity and document transfer of custody. The new purchasing process is refined and accelerated by Friday, but not including the actual fill, still takes nearly four times longer than the prior digital system. Then starting Monday—and continuing for the next four months—the number of tankers arriving to load fuel far exceeds any prior record. Increased demand with slower supply is the wrong direction.

On Monday mid-morning there is a line of tanker trucks extending two blocks in front of the gate to the tollway access road, then nearly to the tollway intersection. Police cruisers wait to give each filled-tanker an escort. There is flow, but profoundly congested. Each small delay accumulates: narrow access, drivers unfamiliar with these racks, on-the-fly transaction processing, a grocery truck stalls on the ramp to the tollway fuel tankers stuck behind. Viewing the backup from his sodden fifth floor conference room Manuel unconsciously, quietly whistles Strozzi's melancholy *Che si può fare* . . . loses the melody and morphs into the Godfather theme.

A GROCERY WAREHOUSE

From the fifth floor Manuel can also see the rooftops of six grocery distribution centers and food manufacturers west and south of the terminal. In a roughly four-mile stretch west of the port, two important fuel terminals and most of Puerto Rico's largest sources of food sit side by side.

On Monday morning, September 25, Esteban is on the roof of his warehouse checking repairs when the backup to the tollway quickly fills the side street in front of the warehouse freight yard. He watches briefly as the knot of traffic tightens, then he descends into the 300,000-plus-square-foot, five-story-high interior.

The leaks were not as bad as feared. Nothing like his neighbor—and competitor—who lost nearly half his roof. The solar panels seem to have added weight and anchoring. An unexpected return on that investment to reduce electrical costs—now powering most of the facility. After a long weekend of cleaning and repairs, the majority of Esteban's 300-some employees are now ready to load trucks.

But most of his biggest customers are not ready to receive deliveries. On Sunday Esteban talked to all but two of his top-twenty customers, together totaling more than 70 percent of sales. None opened today and many do not expect to open until Wednesday, one week after landfall, or even Thursday. The bigger the business, the more complicated the business.

Still, Esteban is making deliveries. Early that Monday morning he had stood on a stack of pallets inside the dock doors.

"Good morning," he shouts above a hundred conversations. "Welcome to the first day of Puerto Rico Rising, with your help. What more can we do for you? The lunch room will open today. I saw many of you bringing your children to our brand-new day-care. Should we keep this going after the crisis?" Applause. "I agree, having the kids here brings an extra joy."

"I am promised the laundromat will be ready by tomorrow after-noon." Louder applause. "Bringing dirty clothes to work is less joyful, but with so much sweat, clean clothes may clear the air." Scattered laughter and groans. "Tell me, tell Carmen," Esteban points to the Vice President of Human Resources, "What do you need? We need you to be ready and able to deliver our food and all that is required for Puerto Rico Rising."

"Today, many of our customers are not yet ready for us. But we have more than 4,000 customers and have not heard from most of them. So, today we are going to them. Every member of the sales team has assignments. We have emptied the cigarette vans and refilled them with packaged goods and water. This morning our direct-store-delivery team will start trying to visit everyone on their routes, find out what they need and bring back orders—food, water all that we have here." Applause and shouts. "Even cigarettes. After Maria, I may start smoking again."

"Please, can I have the sales and DSD teams come forward." Nearly fifty men and women gather between the crowd and Esteban. "You will be our cavalry, our surveillance team, our James Bonds, our sentinels. You will hear what is needed, tell us, and we will be creative and flexible in doing anything possible to deliver all that we have."

"Here is what I expect, it will be slow today and tomorrow. Wednesday the push will begin and by Friday we will be delivering more than ever before. There will be a week of delayed demand to catch up and a huge amount of hoarding. This will continue until the grid comes back, which the Governor says will be by Christmas. If he is right the celebration will continue with super high demand into Three Kings and the eight days beyond."

"Until communications systems are back online, discovering and delivering what is needed will take all the creativity and flexibility we can muster. We do not usually deliver SPAM and SpaghettiOs in the cigarette vans. Not nearly the usual margins. We will have many more custom partial pallets. Much more back-breaking, time-consuming work on the dock. Accounts receivable will be crazy. But your contribution has never been more important. Never have so many depended on you."

"Thank you for your commitment. On Friday morning so many of you were already here. More arrived Saturday and yesterday. Now look at us. We are ready to feed our neighbors. We are ready to do our part in Puerto Rico Rising."

He raises an open palm signaling confidence. More applause, more shouts, Esteban steps down from the pallets and shakes hands with each DSD driver and salesperson. As Esteban departs the crowd for his office on the second floor, the procurement manager joins him on the stairs.

"Regarding that three-month demand surge, I'm not getting boarding numbers for orders out of Jacksonville," she says.

"Tell me more."

"Typically, when we have a container ready the carrier provides a boarding number that tells us roughly when our load will be put on the barge. Our orders are being received, but no boarding number. I expect carriers are keeping their options open."

"No more first come, first serve. Someone setting priorities?"

As they walk past the pop-up daycare, the sharp blast of a trumpet causes both executives to flinch, then laugh.

"That's my guess. Not enough space to push all that's being pulled."

"At least not out of JAX. . . . So, for product shipped from the mainland, keep our orders in the queue, nag as necessary. But if you reach double historical thresholds, pull back. Let's look for comparable products from outside the Jones Act noose. What can we get from Canada, Mexico, Spain or any non-US lanes? Can shippers give us delivery commitments? Perfect opportunity for low-volume international SKUs to claim market share. Worth pushing, might even be true. What can we flex from non-US suppliers to counter congestion and delay in mainland channels? Can we get an actual flow heading our way? See what you can find. Then let's place right-sized, right-*timed* non-US orders."

In the second-floor hallway Esteban is stopped by the logistics coordinator. "Raffie says the port is a black hole. Hacienda still has it basically locked-down. Nothing but FEMA freight getting out. Problem is, he needs to give his guys some work or he'll lose them."

"I want trucks making deliveries too, but I don't want to burn precious diesel sending them *into* a black hole," Esteban answers. "We should have a better idea once the sales team and DSD comes back tonight. I'm fairly sure we will be shipping to Econo, Pueblo, SuperMax, and other big boys by Wednesday or Thursday. What do we need to do to fill the gap?"

"I don't know. I'll ask."

"What I need from you and Raffie is a specific date when our 6,000 gallons of diesel is gone, if my estimate of three-times-normal volume starts on Wednesday, that's a problem I need to solve early. Carmen is pushing to provide fuel to employees. What are the trade-offs?"

Esteban's smartphone buzzes. It's a customer he did not contact on Sunday. "Angel, thank God, so good to hear from you. How are you?"

"Parked beside Route One north of Caguas," Angel answers. "Surrounded by a congregation of three hundred praying aloud to a working tower." It is not a clear signal. "But, anyway, we are open my friend. Eight of nine stores are selling, and we want to see your trucks. Humacao is under water. All the rest are wounded but working, even Yabucoa."

"I'm amazed. When I couldn't raise you yesterday, I assumed the worst."

"Me too. But we're open and crowds are crazy. You got any ice."

Before Esteban can answer no, the line goes dead. He tries calling back. Overloaded.

Rene is one of the cavalry. His horse is a less-than-one-year-old Ford cargo van. He is its first driver. This is his first full-time job and he has been given a low-risk route on the periphery of southern Bayamon, Toa Alta, and into the mountains. He has several dozen low-volume customers, most of whom he sees barely once each month, mostly selling five leading brands of cigarettes, but also watching for any lack of company product on the shelves. Not exactly the job he had in mind when he graduated from university. But it is a start with a good company. He was a marketing intern when his boss recommended him.

For this reconnaissance mission Rene dispenses with his usual circuitous drive staying close to the four lanes of 167 and the 5. This is where most of his slightly bigger customers are clustered.

He wonders if the bridge crossing the Rio Plata has survived. Two of its four lanes have been closed since it was discovered to have been badly built. If he can cross the bridge, he will go as far as the cash-and-carry in Naranjito.

By his third stop Rene has unloaded all the canned goods and water. He considers going back to the centro for more, but really wants to try for Naranjito. He takes orders from six more customers before turning onto the 5 and descending toward the river. The bridge still stands, spanning 2,100 feet, suspended by ninety-six cables fanning from two diamond-shaped towers. It is a gorgeous structure compromised by substandard concrete and an "irregular gradient" producing dangerous vibrations if there is too much weight and speed.

Rene visits three more customers before reaching Naranjito town on the steep slopes above and along the usually narrow Guadiana River. The density of structures protruding from the hills, now denuded of leaves, surprises him. The stream curling through the narrow valley has returned to its typical two feet of flow. Just five days before it was 17 feet high, discharging more than 10,000 cubic feet per second. The homes and shops once gathered along its muddy banks are splintered, buried, or absent. Driving his now-empty cargo van into this implosion of suffering prompts a surge of emotion Rene had not previously allowed.

Fortunately, the main road and the cash-and-carry are well above remnants of flooding. The store is open. Pushing through the crowd of customers, Rene looks for the owner or a manager. The office is locked. Only busy cashiers and three cops at the front end. He finally finds both men and an electrician at the rear of the building nursing one of four generators.

"Rene! Surprised to see you. So, the roads are clear to Bayamon!"

"Clear enough, Gabe, clear enough," Rene shakes each grocer's hand. The electrician remains crouched and peering into the moribund machine.

"What did you bring me. Something more than cigs I hope."

"Not even that, I was empty before crossing the Plata. But I want to get you a truck tomorrow, now I know you're open."

"Two trucks," the manager injects. "Customers are like locusts, eating everything."

"Give Rene our priorities. See how much he can get us tomorrow. How often can you get us trucks?" the sixty-something owner asks the beardless kid from San Juan.

"I don't know," Rene says looking down. "Esteban sent us to find out what was open, where we can go. But I have an idea that might increase your chances."

"I'm all ears."

"I've got forty customers scattered between here, Corozal Comerio, places trucks will not go anytime soon. How about if we load you up and you hand off to these guys, if they can get to you?"

"I'm cash-and-carry. That's what I do."

"Yes, but we deliver these pallets without charging you, reserved for our smaller clients, labeled for each, and charge them directly. If you take their order . . . I'll give you my commission."

"I get it. You're making me a cross-dock. Good idea." The older man pauses, seeming to consider the piles of debris pushed up the hill behind his store. The generator growls briefly, then stops. The electrician curses. "So, you keep your commission," Gabe replies. "But tell Esteban if he gets me at least one truck five days a week, I'll buy him dinner. If I take orders for your customers, he owes me the best brandy you guys carry."

Rene shakes hands with the older man, not sure if he can pull off this brainstorm when he gets back to the centro. He was expecting—dreading—more of a negotiation. Rene is sweating so much that after their handshake Gabe wipes his hand on his already damp shirt.

The next day, Tuesday, two trucks are unloading at the Naranjito cash-and-carry just as Jaime arrives. On Wednesday most of Esteban's customers are ready to reopen. On Thursday Esteban ships seven times normal volume. The outbound flow continues at roughly three times year-prior until mid-December.

By the middle of October fuel is flowing where needed. With sufficient fuel, emergency generators produce electricity and trucks with truckers deliver. Esteban can focus on sourcing, shipping, and resupply.

DARK MOON AND FULL MOON

When his parents brought Jaime back to live in Puerto Rico, the navy brat probably spoke Italian better than Spanish. But even his English was only competent. "Sin musica" (no music) was Nana's complaint regarding her grandson's speaking. So, she became even more musical, quoting Cervantes, reading aloud poetry, and composing Taino myths she would regularly retell in her most elaborate Jibaro dialect. In Vietnam, Desert Storm, and Afghanistan Jaime recalled one particular story to settle—and refocus—his mind and mood.

In the beginning—before Yucajú made humans, before Guabancex remade herself, even before Atabey emerged—there was flow. Those that do not know claim Atabey was self-generating. She is, instead, a specific swirl of primordial flow, giving all within her reach direction and recurrence.

Observing with care, we can see—or otherwise sense—Atabey's touch. Before she mated with Ocean, his waters were still and stagnant. The earth was drenched in stench of stasis. But unfolding within her, Ocean breathed deep and the rhythm of tides began. From the coupling of Atabey and Ocean came contentious twins, each source and creature of powerful flow.

Those that do not know say that Yucajú, the laughing twin, created the moon. Instead, sun, moon and heavens were already flowing, their presence obscured in darkness. Light flowed from conflict between the sacred siblings. Sparks of early friction scattered widely, illuminating the stars. One battle became so heated Yucajú used it to ignite the sun. He intended likewise to set the moon on fire. But Guabancex withdrew rather than risk the night. The somber twin loves twilight and darkness.

In the light of the sun we can know the moon and watch its role in the universal flow. We can see tides rise as our moon swells. With care, we can even discern that we *cannot* see. In each cycle of time there are two tides much higher than others. We are not surprised when the moon is full to find tides pulled higher and higher. We are not surprised when a young man is pulled to a young woman. We are not surprised when the tradesman is pulled to a wealthy market. We are not surprised by the pull of abundance.

But those who know the sea—and those who know the flow— also know that tides are pulled to absence, as the absent beloved pulls the lover, as a poor harvest pulls the hungry. Whenever sun, earth, and moon align—in light or darkness—tides pull higher—as high for the dark moon we cannot see, as for the full moon that gives us joy.

* * *

Then Nana would always ask the twelve-, thirteen-, fourteen-year-old Jaime, "What is the moral of my story?"

5

The General, Juniper, and Jesse: Outsiders

■ ■ ■

We do what we can. We push on . . . as best we can. It isn't for long, you know. Time keeps going by. You'll be surprised at the way time passes.

—Thornton Wilder

THERE WERE, OF COURSE, plans in place for a major hurricane to hit Puerto Rico. These plans—developed and refined through intergovernmental and interagency processes—anticipated that up to 20 percent of the population would need direct emergency assistance. The plans targeted provision of 1.5 million meals per day.

As with most emergency management plans, these estimates were based on prior disasters. Hurricane Georges's hit on Puerto Rico was the core of most planning assumptions. Georges made landfall near Yabucoa on September 21, 1998, as a strong CAT-3. It dropped huge amounts of rain on the central highlands. There was serious disruption of electrical and water systems. All in all, similar force on target.

Georges happened to exit the island going southwest instead of northwest. In 1998 there were about 400,000 more residents of Puerto Rico than in 2017. Cell phones and online systems were not widely used. Another difference: while over 96 percent of Puerto Ricans lost grid connections in Georges, almost all were reconnected within nine weeks. It took nearly nine months to make full repairs for Maria.

Delivering one million–plus meals per day into any disaster zone is an extraordinary feat. Grubhub is arguably the US market leader in meal delivery. It does not operate everywhere. But by carefully developing its sourcing and delivery networks for each market where it chooses to operate, the company delivered more than 400,000 meals per day in 2017.

In the United States the average household of 2.5 persons spends about 6 percent of its income on grocery purchases, roughly $4,200 per year (another 5 percent is spent on non-grocery food). The average traditional grocery store in the United States has annual sales of about $17 million. If the average grocery store supplies two of three meals consumed, this suggests that the average grocery store serves roughly twenty thousand meals per day to four thousand households. As these comparisons suggest, having a realistic plan and the actual capability to deliver one million meals per day into a disaster zone—with little or no on-the-ground operations the day before—is an amazing and even mind-staggering ambition.

Disaster response in the United States is primarily a state and local function. But since the 1980s there has been an increasing federal role in supporting emergency preparedness, planning, mitigation, response, and recovery. Since 2005's Hurricane Katrina there has been a particular effort to enhance federal abilities to support state and local efforts to hydrate, feed, and shelter survivors of the worst extreme events.

Harvey hit Houston. Then Irma swept USVI, Puerto Rico, and Florida. Maria mauled Puerto Rico and more of the Virgin Islands. In each of these cases the federal government moved mountains of material toward anticipated need. Even before Maria hit, it had been a month when huge volumes of water, food, fuel, and other resources had been moved where supplies were uncertain or absent.

Hurricane Maria's eye was eight hours crossing the island, wind and rain preceding and following. On Thursday, September 21, assessments begin: The electric grid is shredded, 60 percent of the water system is dry or dangerous, the governor is unable to contact several mayors, two thousand are rescued from flooding in one San Juan suburb, only seven hospitals have backup power and are able to fully operate, city streets are rivers, pastures and fields appear as lakes, bridges are gone, and roads have been washed away.

For more than ten years Puerto Rico has been in crisis. Economic growth has been negative or stagnant. Public debt has skyrocketed. Population is declining. In May 2017 most legal authority for the Commonwealth's finances is transferred from elected officials to a congressionally appointed board and a federal judge sitting in New York.

In a nonbinding June referendum, an underwhelming number of Puerto Rico voters overwhelmingly call for the Commonwealth to become the 51st State. But political processes seem to have become absurdist theater. In any case, results are ignored or dismissed. Creditors continue to squeeze. More employers depart.

Staggering under $9 billion in debt the electric utility takes bankruptcy. On Sunday, July 2, 2017, in an urgent filing with the US District Court for Puerto Rico, the Financial Oversight and Management Board seeks court permission to keep in place the Fuel Supply Contract on which the power grid depends; otherwise the utility's electric generation would have been disrupted weeks before Maria's destructive collision.

Post-Maria the absence of power is both painfully explicit and insidious. In the three days immediately after the storm's passing, the governor and others are in the streets delivering supplies and wanting to bring hope. Clearing roads begins. Emergency generators are started. Water lines are flushed. But by Monday, September 25, there is little cause for confidence and widespread evidence of the contrary. Not much is moving. Even the floods have dissipated, leaving a toxic torpor of standing water, draining too slowly. Everything stinks.

That Monday the government of Puerto Rico sends the FEMA Regional Office in New York an official form, each section completed with handwriting. In less than two pages, the Resource Request Form (RRF) identifies supplies that the Government of Puerto Rico perceives are urgently needed to directly support survivors of Hurricane Maria. The request includes 500,000 jerricans, 1 million mosquito nets, 25 million baby diapers, 168 million liters of water, and much more. At the top of the list is a request—a need projection—for 346,500,000 meals for the next three months, an average of six million meals per day. The preexisting plans for Puerto Rico call for a mind-staggering

1.5 million meals per day. What does four times mind-staggering do to decision-making? How does this perceived need impact practical ability to serve those in real need? This is Caorao calling on his conch.

DISCERNING REALITY

The General heard about the RRF before he saw it. His guy at the Joint Field Office in San Juan called him. He knows what is in stock, already pushing toward JAX for San Juan. Even before Harvey and Irma, this size demand would have far exceeded supply. Well before the form arrives in New York, the General has procurement reaching out to vendors on what new production can deliver and when. The early answers are not reassuring.

The General is no longer in the army. He is now at FEMA. It's a good fit. The General is an engineer; he builds and repairs. That's what he did with the army after Hurricane Rita hit Texas. After shock and awe hit Baghdad, he contributed to rebuilding that electric grid. As his career has evolved, the General has also engineered how large organizations learn and can help individuals learn. He keeps an Isaac Asimov quote on his desktop, "Science can amuse and fascinate us all, but it is engineering that changes the world." The General also knows that change is difficult, even dangerous. Patience, humility, and do-no-harm are practical virtues.

Toward the close of the Tuesday early-morning logistics update, the General asks a colleague, "Find Dr. Juniper, I'd like to talk with her." He knows it will take a while; Kathy Juniper is famous for constant travel and never answering the phone. But she does watch e-mail closely. An hour later the General and Juniper are in conversation.

The General finishes his problem summary, "So, sometime in the next two or three weeks we can probably gin-up 1.5 to, maybe, 2 million meals per day, but that leaves a huge delta now, even for infinity and beyond."

"I thought the port had reopened," Juniper replies.

"Reopened Saturday. Pretty good shape. Several carriers have already unloaded. Big problem is too many containers on the dock and not enough drayage."

"So, JAX to San Juan flow is resumed, but something is impeding flow from docks into the distribution network?" she asks.

"JAX to San Juan is going, yes," the General responds carefully. "I don't know how to accurately characterize the links between supply and demand. We've got tons of debris on the roads, very spotty telecoms, fuel distribution is all screwed up, truckers are also survivors. . . . You know the drill, every kind of impediment." Dr. Juniper is a supply chain management professor who specializes in humanitarian logistics.

"All these impediments are true for your relief channel too, aren't they?"

"Yes and no. At our current through-put we can deliver in a timely way to the twelve staging areas the Commonwealth has set up. I don't know what happens if we multiply current flow by ten."

"But that's not going to happen because you can't source 10x. You probably can't source more than 2x." It is an accurate statement, not a question, and the General does not respond. Juniper continues, "What does the Governor know that we don't know? Why has he asked for such a huge influx of expensive, warrior-feeding MREs? From everything I can see and everything you are telling me, the preexisting supply network for feeding the people of Puerto Rico has survived this event. Everything needed to feed the survivors is already on the island and more is already on the way, even piling up on the docks. But he seems to assume this is not true. He seems to assume the preexisting supply chain has collapsed and must be at least 80 percent replaced. What does he know that we don't know?"

"He knows much more than I do, sixteen hundred miles away. Not much is moving. If I were Governor Roselló that would worry me too," the General offers.

"Shoving more stuff into a congested system is not the way to solve a problem with movement. This is an issue of velocity, not volume. The volume is there. How do we reestablish velocity?"

"Truckers, trucks, and fuel," the General answers. A onetime mentor had preferred the acronym for Warriors, Trucks, and Fuel. This is a different time, place, and audience.

"I think so too," Juniper replies. "Find your truckers. Find out what is in their way. Remove it. Then you will be well on the way to solving your problem. And General, I don't need to tell you: that request for six million meals may be your most serious threat. Trying to deliver that is going to distract from any real solution. If, in the unlikely event, you eventually source six million meals per day, actual delivery would destroy the preexisting supply chain and seriously delay economic recovery. Those six million meals are charged bombs. Handle very carefully."

"I hear you," the General responds, "but we've got 3.4 million people with no grid, almost no communications, sometimes no water or roof. Maybe the governor and his guys have over-estimated, but given the distance, time-to-recover, potential demand, even time-to-survive, I've got to pull and push everything we can just in case they're right."

"Understood. But if I were you, while your procurement team is trying to source those MREs, I would target all your guys in Puerto Rico on unleashing truckers and releasing food already in the network. Get that done and the six million meal requirement should quickly—deservedly—disappear."

Dr. Juniper asks some data-dependent questions. A staff member on the line with the General promises to provide answers. Juniper promises one of her infamous e-mails. The call ends and the General says to the staffer, "She didn't tell us anything we didn't know, but it helps to hear an outsider confirm. Always good to red-team something this important."

The push for increased supply annoys Juniper. It is a recurring problem in extreme events. The stubborn failure to recognize the predictable ebb and flow of everyday networks is confounding. Her apparent inability to effectively articulate such essential realities is what annoys her most.

FLOW AS FUNDAMENTAL

There are needs or wants expressed as demand. There are different modes of expressing demand. The people, places, and modes expressing demand can usually be known.

There are almost always multiple sources and even sorts of supply. Each source has certain capacity. Different sources are more or less proximate to the expressions of demand.

There are different modalities for moving supply to demand. Each of these modalities have different capacities. Depending on distance, modality, and expression of demand, there will be different routes and therefore time frames and therefore costs for fulfilling demand.

Where there is significant population density and some level of affluence, there will be well-established networks of demand and supply that deliver volume and velocity reasonably matching prior patterns of demand. When the largest populations are involved, *only* this preexisting network has the capacity to deliver the volume and velocity required.

The specific details of fulfilling individual demands across these networks can be quite complex. Amazon, Walmart, Alibaba, Fedex, Grubhub, Uber, and many others are constantly creating new angles and arcs. Friction and surprise are perpetual. But at scale this is a simple model: nodes, links, and flow. Strategic capacities and critical impediments can be known. Perhaps as important, nonproblems can be quickly precluded. At least, this is Kathy Juniper's experience.

* * *

As she begins to research the demand network for food in Puerto Rico, Juniper is self-aware that being annoyed is currently her strongest motivation. There is some vague compassion for victims and survivors. But as happy as she will be to speed and direct food to those in need, at this moment she is even more interested in demonstrating how networks do and do not behave.

She gives her grad assistants some research assignments and self-queries Census, Department of Agriculture, and Department of Transportation databases. Puerto Rico is not always included in federal data gathering, another annoyance. On Wednesday morning she is teaching, not until Wednesday afternoon, September 27, is there some sustained quiet to gather her thoughts. Juniper pounds out some notes. Plenty seems tentative. Nothing strikes her as urgent. She wants to reread and rethink fresh, so she puts the draft aside for Thursday morning. This also allows her to read overnight updates. On September 28 at about 7:00 AM Eastern she presses send.

General:

I assess Puerto Rico has turned an important corner in the last 24 hours. The evidence is mostly related to how grocery stores and pharmacies are reopening, continuing to operate, and being resupplied. I don't have enough visibility to know precisely how this is happening, but I am able to see outcomes . . . especially in metro-San Juan.

Fuel distribution/delivery has been the toughest knot in this crisis. It is now my judgment that the curfew was having a significant—to me, surprising—influence on fuel distribution. (Talk about hidden dependencies.) Once the curfew was clarified to allow night-time deliveries and refueling . . . once retail restrictions on gas purchases were made illegal . . . once the curfew itself was reduced . . . and once grocers and pharmacists recognized the new reality and started putting together ad-hoc solutions, the system adapted more quickly, creatively, and effectively than I would have predicted.

There are reasons to think the evolving distribution/delivery system will continue to strengthen. If this assessment is accurate, it is another example of how almost any arbitrary measure that restricts expression of demand or response by supply creates wicked unintended consequences.

As noted, I am more confident that this description of reality is accurate for the San Juan metro area than elsewhere. I am sure it does NOT accurately describe the reality for many more isolated communities, and even in the metro area there are neighborhoods in desperate need.

About two-thirds of the Puerto Rican population is in the metro-area. For this population:

1. IF they've got some cash, and
2. IF they can get to a grocery store

Then they are increasingly able to get what they want. The big retailers are paying what they need to pay to get the diesel and drivers to keep them in business. The Big Box players are probably not hurting too much. They usually have much higher volumes and their volumes are probably even better now. But many smaller stores may be losing marginally more money every hour they stay open . . . so worth watching your flank if they start peeling off, at least with hours-of-service.

Still, right now more than half of the population of metro San Juan is almost certainly back on the "regular" economy . . . with a serving of "black market" on the side. In the so-called San Juan Urban Zone based on the income structure, I would guess about half the population has cash or can get cash. . .

But if either 1 or 2 is not true, then these survivors are mostly dependent on you and yours. Same is true for a bigger proportion of the non-metro population.

Keeping cash in the hands of everyone possible will reduce demand for your direct feeding. Keeping the grocery sector operating will reduce dependence on FEMA, etc., and should speed recovery. And cash or pseudo-cash should be easier/cheaper to move than shelf-stable-meals.

Food vouchers are one example of pseudo-cash, but I don't perceive these are currently in your inventory of options. In Puerto Rico the single most important example of pseudo-cash is the EBT card administered by the Department of the Family known as the Nutrition Assistance Program or, in Spanish, the PAN card.

- 1,244,440 Puerto Ricans can use this EBT card to purchase food.
- For 935,039 individuals PAN is their only source of income.
- Of 2,885 establishments authorized to receive the card, 660 are currently operating and able to conduct transactions with the card. Disruption of the telecommunications system has disallowed using the card at many authorized establishments that are open.

PAN card beneficiaries are a high proportion of the most vulnerable on the island. This population is probably the group that is—or could become—most dependent on the FEMA supply chain if they are unable to utilize the card to otherwise access food.

Keeping cash or pseudo-cash flowing and restoring telecoms links for Electronic Data Transfer—especially the PAN card—are your most effective strategic interventions for feeding survivors.

Even if my assessment is accurate and holds, there are plenty of treacherous problems ahead. As my curfew confession indicates, the most important issues are probably beyond my ability to predict. But over the last 24 hours the pilings being driven into what seemed endless mud, hit something solid. I understand FEMA's lift is still building. I think FEMAs distribution will now be able to gap-fill instead of attempting to replace the entire food supply chain. This is difficult but possible. Replacement was and remains impossible.

I welcome your questions and push-backs. Kathy Juniper

The first draft had been written for herself, reducing thoughts to paper for her own self-critique. The second draft had been written for the General. The final draft had been written for whomever the General decided should receive a forward. Juniper recognized that anything this long was seldom read in any bureaucracy, and even less in the middle of a crisis. But she considered the situation—both problem and opportunity—to be worth 750 words. Besides, this sort of analysis was her only real value add. She now felt less annoyed.

* * *

After reading twice, the General leans back in his chair. It makes sense. He is sure Juniper's strategic insight is mostly correct. But, not for the first time, he is trapped inside one more Schlieffen Plan: A recurring constraint in large organizations. Once a massive mobilization has begun it is very difficult to stop. Timetables tend to takeover. To stop or slow action, once started, creates other sources of risk. Even a miniscule risk that stopping is the wrong choice trumps otherwise obvious benefits. The General expects this is another of those cognitive biases humanity needs to outgrow, but that is not—at this particular moment in time—his personal role.

TAKING ACTION

Jesse is in his early thirties, single, native of New England, Spanish-speaking. He has been in Panama watching the Atlantic spawn one after another tropical cyclone. The hard hit on Puerto Rico gets him on a plane.

He is a social-capitalist, cross-cultural-entrepreneur, a constructive trouble-maker, a contemporary incarnation of a recurring American archetype: The Problem-Solver. His style and stance are rooted in the context of his generation, a cocktail of post-industrial individualism, improvised community, taken-for-granted technological acumen, and a military-special-operations ethic reflecting America's Longest War. Previously he has worked in hurricane response in the Philippines and Haiti.

On September 26 Jesse arrives in San Juan. He has never been in Puerto Rico before, but friends-of-friends and other network connections provide a quick orientation and help him with logistics.

The most obvious disconnection—and, therefore, potentially the greatest need—is in the mountainous interior. Jesse drives to Oracorvis and Barranquitas (where Jaime's microwave tower still lies flat on the ground). By the second day walking around and listening—and having the language skills to hear—Jesse recognizes that to serve the largest number in the most impactful way, there is an urgent need to reconnect the EBT transaction terminals in the grocery stores and other food shops.

But while the constraint on flow can be quickly diagnosed, Jesse does not have the personal expertise to prescribe a solution. He reaches out to others who know telecommunications. There are several technological options, and after a few hours' troubleshooting they settle on lightweight, so-called satellite-in-the-box connections for each store.

Demand is known. Supply is known. One important impediment in between has been identified. A solution is now known. An entrepreneur is predisposed to quickly fill the gap. The contemporary Internet-era entrepreneur is especially inclined to fill the gap at scale. Jesse and his colleagues engage the Joint Field Office, Commonwealth officials, anyone they can find with funding or functional reach to solve the problem at scale. There are good discussions. But no movement. This is a network interval that falls outside the current disaster response system. No one is pre-authorized. Everyone says, good idea. No one says: here's the money, here's the order, here's the team with technology to deploy.

In the free market there are always barriers to entry, always impediments to adopting new ideas. But there are also systemic work-arounds and innovative structures to speed solutions to market. Venture capital, innovation hubs, and Supply Chain Rapid Restoration Teams are not (yet?) part of the disaster preparedness, response and recovery landscape.

At the beginning of the third week, frustrated by delay, Jesse orders eight solution sets and puts the $30,000 purchase on his personal credit card. By the end of that week, the technology is installed by a for-profit social benefit company's disaster support team. As far as Jesse knows, no one replicated Pepita's dancing. But there was plenty of excitement in each store when the PAN card could be used again.

It was a quick, easy, and comparatively cheap solution. For each of the eight grocery stores receiving the satellite technology—and for the communities they served—the restoration of EBT transactions was transformational. But only eight?

Dr. Juniper could identify 2,200 locations needing reconnection. Not all of these locations survived the hurricane. Not all were being resupplied. But it is worth noting that providing a satellite-in-a-box for 2,000 preexisting food retailers authorized to receive PAN cards would have involved a direct cost—not including installation—of about $7 million. It has been conservatively estimated that FEMA spent $300 million for food in Puerto Rico—not including transportation, distribution, and storage costs. Could have satcom installations paid for themselves?

LIMITS OF MASS MOBILIZATION

The federal government was entirely prepared to spend even more. Once rolling, mass mobilization can easily become self-reinforcing.

Between September 24 and October 12 roughly 8.4 million shelf-stable meals are delivered to staging areas in Puerto Rico. In most cases, these resources are then distributed to survivors by National Guard or municipal personnel.

Shelf-stable meals and snacks feed a body, but they can soon seem to drain the soul. The second week in October, a FEMA and US Army plan circulates for delivering hot meals and fresh food. The Concept of Operations aims to provide 2.6 million daily fresh and hot meals by the end of the third week in October, sustain this through mid-November, and then wind-down mass care operations by early December.

As this volume and timetable reflects, by early October it is recognized that a target of six million meals per day until Christmas is no longer necessary. Despite this recognition, the new plan is mostly indiscriminate in terms of where meals will be distributed.

Kathy Juniper is opposed to the Mass Care Feeding Concept. In teleconferences she calls it the Garrison Feeding Concept. In her judgment, by mid-October the grocery supply chain is doing a good job in most places. There are some specific gaps and that is where army and other resources should be targeted. Island-wide feeding is not needed and could threaten the grocery demand-and-supply network. Very annoying.

While no one has said so and she cannot prove it, Juniper suspects the lack of targeting in the plan reflects the political need to provide equivalent services to each municipality rather than any assessment of actual population needs. Even more annoying.

Juniper argues on phone calls and makes a trip to Washington and sends long e-mails. But by October 14 it appears the plan will be implemented. She perceives the General is also a skeptic, so as a last try to blow a referee whistle, Juniper sends an e-mail that she hopes he might forward to endorse mitigation-of-harm.

General:

It looks like the Mass Care Feeding Concept is pretty much locked and in the process of being loaded. This will feed a million-plus Puerto Ricans.

As supply chain geeks and systems thinkers, our parallel concern is how to mitigate any negative impact on the commercial supply chain, in terms of the private sector's current role in feeding the population, in terms of expediting and sustaining economic recovery, and in terms of the eventual stand-down of "relief" operations. Personally, I am concerned that once implemented the "market" will require the Mass Care Feeding effort be sustained for much longer than the current concept anticipates.

I suggest that as much as possible—both substantively and in terms of a communications strategy—the Mass Care Feeding Concept should be focused on geographies where there is not a proximate grocery store AND there is no current capability to make transactions using the PAN card (AKA the Family Card or NAP Card). Where a grocery store is able to transact with a PAN Card, please give them the space to keep operating. Otherwise Mass Care Feeding will kill them.

Further, where there is no current capability to make transactions using the PAN card but there is a grocery store open, I suggest making the grocery store the Point-of-Distribution for the Mass Care Feeding Concept, explicitly pending reestablishment of PAN capabilities.

Purchase product from the stores and pay the stores to provide the fresh food and hot meals. In this way you are pointing to the long-term and providing a much more sustainable, much more customer-friendly solution . . . while also demonstrating a commitment to addressing the current crisis.

I understand many grocery stores have begun operating barbecues and other expanded prepared foods operations precisely because many customers do not have the ability to cook at home. But some grocery stores have not had the capital or product to provide this service. Some will welcome having FEMA help get it started . . . and once the PAN card (and other bank cards) can be used, the mass care feeding "function" at the store becomes commercially sustainable.

Further, I suggest standing-up a strike team to do everything possible to reconnect grocery stores to be able to receive PAN transactions. As this becomes possible draw-down Mass Care Feeding in coordination with each grocery store.

Finally—at least for tonight—a comment on fuel and distribution: For many grocery stores—especially outside the more prosperous neighborhoods, the combination of the price freezes and the increased cost of trucking and cost of fuel for back-up generation is potentially causing some of them to actually lose money each day they operate. This is fundamentally undermining their ability to order goods and simply keep the doors open. Especially if these stores cannot receive PAN card transactions, many of these operations are unsustainable in the very near-term. If these outlets close, the dependence on the FEMA/ARNORTH supply chain will be wired-in. And the recovery will be that much tougher. Any creativity that can be deployed to support grocery stores as PODs will significantly enhance the recovery. In every case try to maximize use of preexisting assets.

No response arrives for this e-mail. Juniper assumes it is another example of her failure to effectively communicate.

* * *

In various forms the mass care feeding mission continues until March 31, 2018. Over sixty million shelf-stable meals are shipped to Puerto Rico; many more meals are provided with FEMA funding.

6

Acaso Una Intención (Perhaps an Intention)

■ ■ ■

We come from a world where we have known incredible standards of excellence, and we dimly remember beauties which we have not seized again; and we go back to that world.

—Thornton Wilder

"NINE SAILINGS—JAX TO San Juan and back—with no break is no fun," Alvarado says. "But no matter how fast we turned, no matter how tight we packed, there was always more."

"I would meet him at the dock," Ana adds, "pick him up and his bag of smelly clothes. He would take a bath. I would put the clothes in the washer. We would take a walk. Then I put the clothes in the dryer. He would fall asleep. Eventually I would join him. *Maybe*, every second trip, there is time for a little *something*, but then it was back to sea."

"There was not any let-up until late January and we're still moving more than we have in years. Absolutely amazing."

"But you're ambivalent. Back on the docks every chance I gave you to brag, you went the other way," Juniper says. Looking at Ana, "Is this just his personality, pushing away praise."

Ana laughs, "No, Paul Alvarado is not a humble man!"

The three are seated inside a seafood restaurant overlooking the big bend of the St. Johns River at Jacksonville. It is over-air-conditioned, but it is too hot and humid outside this mid-May day. Captain Alvarado takes a long sip of his Manhattan.

"I'm proud of what we did. We kept that island alive. USVI too, by the way. But . . . but. . . ," he pauses, gathering his thoughts.

Juniper starts to offer something, but literally bites her tongue instead.

"It was probably the biggest Beer Game ever played. Think about it, the system that had been feeding three-and-half million people on September 19 was mostly still in place after landfall. All the food that was in the pre-existing pipeline was still flowing. Nothing had happened here or at San Juan to change anything fundamental. People just got nervous. Grid's out. Communication's spotty. Lots of stores aren't open. Truckers are absent. So, yeah, folks understandably get nervous. At the port and on the liners and on order into JAX, we've got everything they need. But they ordered more and more. Demand was fear-driven. New sources of demand—like FEMA, Commonwealth, and Mayors—ordered a lot more on top of the plus-up already flowing. So, we're trying to push a lot more down essentially the same pipe. Was there displacement? Hell, yes. Was there a need for much better distribution of stuff on the island? Absolutely. Was more stuff needed? Not really and handling all the unneeded stuff just complicated solving the distribution problem. There were serious communications and distribution problems, misperceived as supply problems."

* * *

Three days later Juniper is seated with Esteban in his office. "When did the demand surge finally plateau?" she asks in English.

"Well, year-to-year we're still above any comparable month for over sixty months. But the pig-in-the-python—the peak—was mid-December."

"What happened, why then?"

"Nothing in our data pops out at me. If you want to dig into the data, I suggest taking a hard look at a potential correlation between bottled-water inventories and retail sales. My impression—just anecdotal—is once stores could keep their water aisles stocked, the anxious buying was seriously curtailed. My guess is that consumers took that as a crucial signal they could be confident of future supplies, for water and more."

"Sounds like a promising graduate thesis," Juniper smiles. "You've shared so many issues and solutions. When you look back now—five months after peak outbound—what do you consider the three most critical challenges?"

"At an April retreat, the executive team asked precisely the same question. In just about three weeks a new hurricane season begins. What did we do right? What do we want to do better? We solved several problems and we don't want to forget what we did right. But there were three persistent sources of pain: First, communicating with our customers. Second, delivering to our customers. Third, receiving timely and assured resupply. I don't know how we achieve the third as long as the Jones Act is in place. Our options are severely limited because of that entirely arbitrary constraint on flow. In terms of delivery, we have made a strategic investment in a strengthened trucking partner. We think this will give us the capacity we need to serve our customers every day and give us much more capacity on bad days. We will still contract with several trucking providers. But on bad days, this primary partner should be able to focus on moving our product to our customers. For communications, we are in focused discussion with each of our top-twenty customers. Each one is unique. Many have already invested in their own solutions. But our shared goal is that no later than June 5 we will have specific bilateral arrangements with each giving us assured voice and EDT connections."

* * *

After lunch Juniper is seated with Manuel and three of his colleagues. "Accounts receivable is still in discussion with several customers. We have still not reconciled our records with their records," Manuel explains. "But I will also add that all of these discussions have, so far, been remarkably cordial. We went through a tough time together. They appreciate we leaned-forward. We understand our on-the-fly transaction system was not perfect."

The CFO—a Sloan School MBA—comments, "Our experience with digital transactions has wider implications. In any high-volume, high-velocity network there are certain functions that optimize throughput and targeting. Four years ago, when our digital system was fully implemented, we felt like we had achieved nirvana. A major source of friction in the network was removed. We more than doubled capacity by changing a non-physical process. Wow. But each time we increase this kind of functional efficiency we introduce risks, some obvious, others less obvious. We thought through many risks, including if we lost digital processing. But not far enough. We did not—I did not—adequately consider what would happen if all our customers lost their digital connections. Instead of just thinking it through, we should have worked it through—functionally tested it—with our customers."

"To that point," Manuel adds, "over the last five years we have spent tens of million updating and consolidating our systems at this terminal, reducing our footprint at two other legacy locations on the island. This has significantly improved our efficiency, safety, and environmental controls. It also concentrates risk . . . for the company . . . and given our important role, for Puerto Rico."

"I feel like there's another paragraph," Juniper says.

"A rather pregnant lacuna, yes." Manuel smiles.

Juniper laughs, "Well, pregnant with what? A bridge or a deeper chasm?"

Manuel cocks his head. "I predict fraternal twins. We want," Manuel extends his arms to the others around the conference table, "to build every bridge that is needed. But we do not have time to build every bridge. We do not have money to build every bridge. We will not see every bridging opportunity. We will unintentionally deepen and widen the chasm."

* * *

The next morning Juniper is scheduled to meet the General's senior guy in Puerto Rico. The JFO still exists but has long since moved from the Convention Center to a suburban office park. Juniper's driver has been there before but was driven by someone else. Now she cannot seem to remember the way. GPS is working again but is obviously misinformed. They call the JFO two times for direction. They are clearly very close but cannot find it. The third time Juniper calls to reschedule. They meet two nights later in the café of her hotel.

"I've been here almost eight months and I know almost nothing about the grocery sector in Puerto Rico. Sad but true."

"Then how did you target food deliveries?" Juniper asks.

"My pull signal was the Commonwealth's resource order. I source and move what is requested to the designated location. This pull signal, I learned, usually reflects directly or indirectly what each mayor perceives is needed or worries might be needed or thinks would be helpful to make available, whether needed or not. How well a mayor understood his or her local food status varied widely. But they certainly understood it better than me."

"When did the food distribution end?"

"March. Ash Wednesday was March 1 this year. Someone said—joking—that Puerto Ricans gave up MREs for Lent. But pull was declining all across February. Do the grocers hate us?"

"Most complain about FEMA's free food," Juniper answers. "But the reality is most of them were selling more than they had in years. One of the grocers told me he perceived customers bought what they wanted from him then stood in line for free food to build up a stock-pile for the next hurricane season."

"That was not the mission assignment," he says shaking his head.

"No, but not a bad plan, all things considered." Juniper continues, "So on this issue, let me try out something on you. I keep hearing the number 60 million tossed-about as the number of meals delivered. Is this accurate?"

"Roughly, give or take a million. We'll eventually peg a number, but with damage, non-delivery, and all the rest, 60 million will do."

"Delivered between September 20, 2017 and March 1, 2018?"

"Well, I think the count actually begins with the Saturday, September 23 shipment and we had some lagging deliveries into late March. But, if you're looking for a calendar count I'd say 160 days would work."

"Great." Juniper opened a notebook and began writing numbers. "So, if we have 3.4 million people on the island, let's say they only get two meals per day. Two shelf-stable meals go a long way in terms of calories. That totals 6.8 million meals needed per day. Multiplied by 160 days that equals 1,088,000,000. Just to make the math easier, how about we say 1 billion meals?"

The FEMA official is quiet.

"So, 60 million is six percent. Right?"

"Well, the 60 million doesn't include all the calories delivered. We had Salvation Army, Red Cross, Jose Andres. . . "

"Maybe 10 percent then?" Juniper suggests.

"Probably not that much more . . . My head wasn't anywhere close."

"It's still a huge number. Absolutely lifesaving in many cases."

* * *

One week later, Jaime is showing off his two "new" preowned generators to Juniper. This is her third trip to Montemayor, but her first conversation with Jaime. He had been away from the store on the prior two visits. Juniper had interviewed Pepita and Carlos with the help of a translator.

"With these two babies I'll have redundant back-up and a combined storage capacity sufficient for almost seven days," he says.

"Did you ever lose power again after that second week?"

"No. By mid-October the fuel supply seemed to be operating normally. It was almost like that outage was the system resetting itself."

"How about grocery deliveries, when did they come back?"

"Depends what you mean by back. Goya, B. Fernandez, Plaza . . . the trucks were back on our route and some kind of schedule after a month more or less. I couldn't always get what I ordered. I still can't get everything I order. But I didn't have to drive to Naranjito every other day."

"Carlos and Pepita seemed to remember the power being reconnected in mid-March?"

"Friday, March 16, 4:30 PM," Jaime says emphatically. "There's a story there. The mainline had been live for several days, but we couldn't convince the utility to reconnect the spur that feeds the pharmacy and me. Remember I was paying about $300 per day for diesel.

Anyway, the pharmacist and I paid two of the linesmen to do some freelance work. Saved us a lot of money."

"So, by the middle of March, you had your point-of-sale terminal back, deliveries are being made, power is back on. Back to normal?"

Jaime does not answer immediately. "No. Normal is gone. Forever. Maria took her away."

After a long beat he continues, "You know, it's okay. I'm opening a second, bigger store. My brother and I are starting a cash-and-carry, really good use for the box truck and if I get enough volume going, I'll be able to lower my retail prices by—maybe—3 percent and still make more profit. But even less time for a cortado. And my wife, she would throw me out if I smoked a cigar. Ha!"

"What's the biggest change?" Juniper asks.

Jamie looks carefully at the Yankee academic before answering, "I know this isn't what you're asking. But the biggest change is that my aunt, Pepita's mom, died in November. She was the last of that generation, the last to live through San Ciprian. She was too young to really remember, but she had lived the stories, and she could retell all my grandmother's stories. Tia Elena was an old woman. But Maria took her too."

Juniper and Jaime avoid eye contact.

He continues, "My grandmother had a favorite story, it's a Taino myth about the tide rising as high for the dark moon as the bright moon. She always asked me for the meaning. I've found or made many different meanings. Today, my meaning is that in good times and bad our most productive response is to find the forward flow and stay with that flow. No matter how frightening or turbulent: don't resist, don't impede, don't be distracted. The flow will go where it will, one way or another. It is our task to bring love into the flow. We can learn to ride the flow where we need to go."

The Moral of Our Story

■ ■ ■

*I am not interested in the ephemeral. . . . I am interested in those
things that repeat and repeat and repeat in the lives of millions.*
—Thornton Wilder, but not from *The Bridge of San Luis Rey*

NANA ASKS JAIME to tell her the moral of the story: Cutting
through all the mythic detail, what does it mean? What are the practical implications? What does the story teach us about living? What
can we do with it?

Jaime, Pepita, Captain Alvarado, Manuel, Esteban, the General,
Juniper, and others are active participants in a very interactive network.
Demand pulls. Desire pulls. Money pulls. Supply is pushed toward
signals of demand. Supply seeks its way, as fluidly as possible, around
impediments and constraints. Flow emerges. Flow tends to persist.

This is how networks behave. Networks flux and fail with infinite
detail, but several general patterns of behavior repeat and repeat and
repeat. Understanding this repetition can be helpful.

GRAPHING MOVEMENT

Networks are structures that reflect relationships and influence
behavior over time and space.

We are someplace; we want to go someplace else. There are
origins. There are destinations. There are potential paths between.

81

Perhaps others join us in wanting to go between here and there. More or less randomly, we explore possible connections between here and there. We find—or make—a preferred path from here to there. If this individual preference comes to be shared by others, shared movement may generate network characteristics (e.g., greater security or convenience) that reinforce preference and attract increasing use.

Once upon random pathways may begin to display preferential consequences, people and products clustering where varied paths intersect. As preferences play out and if cluster size grows, networks often accumulate functional characteristics that facilitate increased volume and movement at greater speeds in specific directions. Building a bridge over a river is a classic example: prior to the bridge, there may be several well-used fords, none of which have any persistent advantage over another. But the comparative advantage offered by the bridge results in many prior paths being mostly abandoned. Recognizing this convergence of travelers, services—such as food and lodging—emerge around the bridge, creating even greater comparative advantage. Overtime preexisting paths at a considerable distance from the bridge tend to turn toward it, and the entire network increases its dependence on the bridge.

For centuries products and people flowed into and out of Puerto Rico from many different places through various ports of embarkation and disembarkation. San Juan has almost always been the principal port. But today San Juan's proportion of flow is much higher than ever before. In 1917, 1,905 vessels carried 5,169,000 gross tons of goods to Puerto Rico. Of these, 720 vessels with 1.9 million tons docked at San Juan.

CONCENTRATION

Today San Juan handles ten times more cargo tons than all other Puerto Rican ports combined. As paths converge around comparative advantage, people and services tend to cluster at places where two or more paths converge. These nodes—the swelling of demand and/or supply at a particular place—often facilitate flow: with markets, financial brokers, transportation resources, and other services to direct and accelerate movement. In some cases, these nodes become population centers, places of wealth, and their own significant sources of demand, all of which reinforce pull across the network toward the node.

As mature, high-volume, high-velocity networks achieve scale, there is a persistent pattern of a small number of nodes taking on an outsized proportion of flow. Over time the flow of people and products self-organizes around a preferred set of paths and nodes. (This is sometimes called percolation, making an analogy to how fluids find/create new nodes within and paths through porous media.) Flow tends to concentrate—we might say, flow tends to pool—in places with a depth of demand and the ability to influence the velocity of flow. At the macro-level great urban concentrations—New York, Los Angeles, Tokyo, Shanghai, and London—both attract and accelerate flow of demand and supply. High-volume demand tends to transform natural percolation into engineered permeation, channeling pathways between nodes across a wide landscape.

Over the past century, dense concentrations of demand—for water, food, fuel, pharmaceuticals, and other essentials and nonessentials—have transformed the physical and functional landscape of traditional demand-and-supply networks (sometimes referenced as "network topology"). This reshaping of landscape and relationships has been especially dramatic over the past fifty years, and the pace of change is likely to persist well into the foreseeable future.

In 1977 six US cities had a population of one million or more. Today ten US cities have a population of more than one million, and there are at least fifty-three US metropolitan areas with more than one million residents. A half-century ago 73 percent of the US population lived in urban areas. Today over 80 percent are urban residents, and the rate of urbanization is increasing. To serve this burgeoning demand, supply chains increasingly self-situate across space and self-synchronize across time to find, create, and reinforce comparative advantages.

The 3.4 million more-affluent-than-average residents of Puerto Rico are a source of considerable network pull. Since 1977 the GDP per capita of Puerto Rico has increased from $13,000 to just over $30,000 (in constant 2010 dollars). The tendency of self-organizing networks to concentrate capacity in a small proportion of nodes has been reflected in the increasing importance ("centrality" to use a network science term) of the Ports of San Juan and Jacksonville.

A host of factors have played out to give preference to the maritime bridge between JAX and San Juan: increased use of shipping containers has given comparative advantage to sea ports that were early adopters of container technology. Container technologies deliver comparative advantage to large-volume shippers that have the financial ability to absorb high fixed costs while slashing variable costs. In the particular case of Puerto Rico, the Jones Act constrains the competitive context and since 2006 the Commonwealth's economic crisis has eliminated weaker shippers. Over time connections other than JAX-San Juan have diminished or been abandoned.

BETWEENNESS

San Juan is open to the world. But San Juan is connected to the world through a rather small set of ports. The value of San Juan's shipments in and out of Jacksonville is greatest by far. Both ports sit *between* enormous pools of supply and demand. Most paths connecting Puerto Rican supply and demand pass through both ports. (There are network science methods for measuring "betweenness.")

Both Jacksonville and San Juan ports demonstrate the seeming structural predisposition of high-volume, high-velocity demand-and-supply networks to assume hourglass characteristics. Large and widely distributed sources of demand pull large and widely distributed sources of supply through very narrow necks (or waists, both terms are used in network science literature). Kaeser M. Sabrin and Constantine Dovrolis, network scientists at Georgia Tech, argue, "The hourglass effect should allow a system to accommodate frequent changes in its sources or targets (i.e., to be able to evolve as the environment changes) because the few modules at the waist 'decouple' the large number of sources from the large number of targets. If there is a change in the inputs (sources), the outputs do not need to be modified as long as the modules at the waist can still function properly. Similarly, if there is need for a new target, it may be much easier (or cheaper) to construct it reusing the modules at the waist rather than directly relying on sources."

Very similar patterns repeat fractal-like across the network. Manuel's fuel terminal becomes a node with a high "betweenness" score. His rack is one of the flexible modules at the waist. Esteban's distribution center is another node with many connections. Esteban and Manuel manage many modularized functions—storage space, docking capabilities, trucking, personnel, capital, and so on—that can be applied to a wide variety of inputs and delivered to many different targets across time and space. Rene's brainstorm to use the Naranjito cash-and-carry as a cross-dock increased its connections. These are connections that can be mathematically measured by "betweenness" and other aspects of "centrality." Jamie's store, generators, trucks, and more are another neck in another hourglass embedded within several other hourglasses, a chorus line of ghostly Russian Matryoshka dolls dancing across the matrix of time and space.

BOTTLENECKS

The neck of a bottle is the neck of a potential—or once-upon-hourglass, now separated from supply or demand. A so-called bottle-neck is either unable to send signals (pour its contents) or receive signals (fill its void). Something is impeding flow.

In Puerto Rico missing bridges were among the most prominent bottlenecks. Debris in the road and long-term loss of traffic lights complicated, redirected, and slowed flow. But the most difficult and recurring bottleneck was probably the loss of digital financial transactions (pure signaling). At the port, fuel racks, gas stations, and grocery checkouts, the bottle could not be fully uncorked. What had been a strong flow was much reduced while digital connections to financial transaction nodes were broken. Once the digital signaling had been restored—Jesse and his friends showed one way to do it—the prior flow came back very strong.

In many ways this entire story has been about finding and fixing bottlenecks. Bottlenecks are a recurring feature of demand-and-supply networks (and other natural and engineered systems). Even on the best days, impediments—accidental, structural, or intentional—emerge across networks.

Bottlenecks do not always emerge at the neck of an hourglass, but when they do, the impediment or separation has amplified effect. This begs the question: given the risk, why do many demand-and-supply networks display hourglass characteristics or, in some cases, actually form hourglass-similar spatial silhouettes?

To persist with the analogy, narrow necks allow bottles to be more accurately poured and assist in keeping unwanted bits out of the pour. Hourglass structures emerge in demand-and-supply networks to facilitate speed and accuracy of flow—and to contain costs. These are essential components of comparative advantage in any real-world network.

The functional modules created to facilitate flow and contain costs at the neck of an hourglass—such as cross-docks, fuel racks, and digital financial transactions—can be effectively deployed across a wide range of sources to serve a diverse set of targets. Entirely novel sources (new products) can often be delivered to previously nonengaged targets (new customers) by the very same modules.

While functionally flexible, this is also an inherently conserving system. Any product or customer that cannot be served by some version of existing functional modules is unlikely to be well served.

Fundamental changes in functional modules can have revolutionary impacts on the entire network. For example, shipping containers have radically transformed the topology of global trade as much as or more than the emergence of steamships. After 5,000-plus years of break-bulk shipping, in the past fifty years, this "new" functional module has increased the speed and accuracy of flow while containing per-unit costs for many sources and targets (as well as contributing to the disappearance of weaker—or just smaller—network participants).

Given the significant network effects of the functions concentrated at the neck of an hourglass structure, it is not surprising that serious consequences unfolded from loss and diminution of module capabilities after Hurricane Maria.

I am surprised the consequences were not much worse. Given the scope and scale of destruction and disruption, I am surprised by the resilience of Puerto Rico's preexisting network of demand and supply.

According to some network scientists, this resilience can also be explained by characteristics of the modularized functions at the neck of an hourglass structure. The structure itself constrains content. Prior to 1760 most hourglasses were two separate bottles connected at their necks by waxed cord, shaped and tightened to regulate flow. The neck of an hourglass is rate-limiting—or, perhaps, more accurately: rate-shaping.

Research from biology, ecology, chemistry, information sciences, and more has found recurring examples of constraints that deconstrain. Organic networks emerging without human intention often assume hourglass structures and feature functional modules performing behaviors analogous to that we see in demand-and-supply networks. The human heart is a set of modularized functions at the neck of the circulatory system's hourglass structure. Cardiac functions constrain, even as they enable, empower, facilitate, and, in these ways, deconstrain human capacity.

ASSESSING AND ADVANCING FITNESS

Alvarado, Jaime, Pepita, Manuel, and Esteban spend most of their lives swimming the flow through and around the neck of their hourglass structures. Like fish in water they are simultaneously expert and oblivious.

The General and Juniper are smart, attentive, experienced, and well intended. They are also far away and not quite sure about the practical implications of what they seem to sense. In approaching network behavior, we can often be like medieval physicians, being aware the heart is important but not knowing how it functions.

Jesse followed the heartbeats and listened and looked for any persistent cardiac arrhythmia. In his own words, he was looking for "basic system failures" that cascade across networks in secondary and tertiary consequences. Find the core cause, fix that, and the related symptoms often disappear. Too often we treat only symptoms. Sometimes our treatments exacerbate the core cause.

Months after Maria, I was talking with a FEMA official who said, "I need some way to get a quick, accurate read on supply chain fitness or non-fitness. I don't need—and don't have time for—an MRI. But I sure need to know about anything that is interfering with breathing and heart-function. Where is the bleeding? Where should I focus? Where am I not needed?"

The toughest need to fill in Puerto Rico was increased fuel distribution capability. Given the ongoing demand for generator fuel, there were not enough tanker trucks and drivers to move products from supply nodes to demand nodes with sufficient frequency. What was not needed in most places was more food. As with fuel, there was a challenge moving food from supply nodes to demand nodes (partially because of the problems with fuel distribution). Unlike fuel, in most places there was no increased *need* for food (demand is a different story). But especially in the first two or three weeks there was no credible process for anyone to know what was happening or not happening.

We now know how to generate meaningful measures. But I am not aware of any sustained and systematic process for monitoring bottlenecks during the next big hurricane (much less the next big earthquake) in Puerto Rico or elsewhere.

Bottlenecks can emerge almost anywhere in a network. But bottlenecks—network aneurysms—involving functions at the neck of an hourglass structure tend to have amplified effects and can cause catastrophic consequences. Recognizing this vulnerability and knowing how to ease the constraints (and reinforce functions that deconstrain) is a very helpful skill. Jaime, Manuel, Esteban, Jesse, and Rene each demonstrated this skill.

Consumer hoarding often follows from congestion, disruption, or destruction at the neck of the hourglass in demand-and-supply networks. Uncertainty of supply tends to unfold as an unsustainable surge in demand, which further disrupts the network. This is the pull of Nana's dark moon of absence. The more quickly darkness becomes light—the sooner that survivors see credible signals the preexisting supply chain is recovering—the sooner hoarding will stop. Rationing is an obvious signal that the supply chain is not functioning near normal. Rationing spurs hoarding. Moreover, in many cases the process of rationing itself disrupts network flow, as it did with retail gasoline and diesel in the first several days after Maria. Once hoarding begins can it easily become a vicious cycle.

In Puerto Rico several factors were crucial to reestablishing network equilibrium after Maria. The persistence of incoming flows of fuel and food was fundamental. If the maritime bridge or the Port of San Juan had been seriously impeded, supply chain recovery would have been much more difficult and could not have been sustained. The hourglass did not shatter.

The General's early delivery of additional fuel tanker trucks was also crucial. This expanded key functions at the neck of the hourglass. Deployment of electrical generators for the water system was as important but would have been meaningless if fuel distribution capabilities had not expanded to fulfill the needs of these and thousands more generators.

As Esteban and Rene, Jaime and Pepita demonstrated, recovering and redirecting functions at the neck of the hourglass—with the grid down and precarious fuel supplies—required significant creativity, persistence, and insider know-how.

Each of these people are deeply connected within their own networks. Manuel and Esteban (and the several real persons on whom these composites are based) participate in huge global webs of supply. Jamie and Pepita operate within a smaller hourglass but have an intimate knowledge of nearly every grain of sand and percolated path between the sands.

The functions of the human heart are regulated by the autonomic nervous system, a largely unconscious set of biochemical feedback mechanisms that connect the heart to many other body functions. Many large networks feature a pattern of complex self-organization and emergent coordination among largely autonomous agents. Such behaviors are labeled stigmergic or sematectonic or as instances of swarm intelligence. Depending on the agents and systems involved, the mechanisms of feedback and coordinated action can be dramatically different. But what these phenomena tend to share is a persistent pattern of signaling that prompts increasingly self-similar and symmetrical behavior by a significant proportion of the autonomous agents and, as a result, observable changes in network behavior.

Especially at scale, complex demand-and-supply networks depend on stigmergic coordination. Days before a hurricane hits the North Carolina coast, thousands of trucks increase deliveries along and through Interstate 95. In the forty-eight hours before the hurricane arrives, thousands of trucks shift west to Interstates 85 and 77. No command is given. No network-wide control exists. But signaling prompts shared action over a wide area.

It is not difficult to find major nodes in demand-and-supply networks. Some links—such as major truck routes, maritime lanes, and pipelines—pop out. With careful observation, hourglass structures can be discerned, especially at obvious pinch points including ports, big bridges, and mountain passes. It requires considerably more care to determine the size, direction, and comparative value of flows between any particular set of nodes. The comparative value of flow-as-flow is still more challenging. Even after recognizing an hourglass, understanding the modularized functions at the neck may require insider knowledge. But all of this can be done well in advance of an extreme event.

Jesse, the General, and Juniper were able to map Puerto Rico's key networks. Our other major characters inhabited and enabled these network elements. But mapping and understanding is not the moral of this story.

Rene decides he wants to get to Naranjito. After arriving in Naranjito, Rene conceives of converting the local cash-and-carry into a regional cross-dock. The owner of the cash-and-carry agrees. Esteban agrees. So, on Tuesday after the storm, Jaime is able to pick up desperately needed groceries on credit with his wholesaler. A flow of supply to Montemayor resumes.

* * *

I hope many readers of *Out of the Whirlwind* will dig out their high school copies of *The Bridge of San Luis Rey*. My story shares intentional examples of self-similarity. There are complementary concepts of reality and the human condition.

In a letter to a friend, Thornton Wilder wrote: "It seems to me that my books are about: what is the worst thing that the world can do to you, and what are the last resources one has to oppose it. In other words: when a human being is made to bear more than human beings can bear—what then?"

Fear and courage. Generosity and hoarding. Clamor for control and outbreaks of collaboration. Bureaucracy and creativity (even by bureaucrats). Insightful thinking and ignorant reactions. Egotistical commands and merciful self-giving. This age-old dialectic of opposites was certainly on display in Puerto Rico after Hurricane Maria.

Wilder explains, *The Bridge of San Luis Rey* "asked the question whether the intention that lies behind love was sufficient to justify the desperation of living."

Love is hard. Not even the strongest human love can guarantee a good result. But is the intention that lies behind love sufficient to justify the desperation of living?

Life was desperate after Maria. Life was desperate in the midst of California wildfires, at Tinian after Yutu, across Tohoku following the Triple Disaster, in the ninth ward after Katrina, in Tacloban after Yolanda, in tsunami after tsunami. . . . Someday life will be desperate in Seattle, San Francisco, Los Angeles, Miami, Memphis, and Houston.

Does intention matter? The human response to Hurricane Maria in Puerto Rico demonstrates that intention can accumulate: as hoarding or giving, in fear—but also in lifesaving care.

Afterword

■ ■ ■

Let me live now, she whispered. Let me begin again.
—Thornton Wilder

THE PATRON SAINT of Puerto Rico is San Juan Bautista, St. John the Baptist, slightly older cousin of Jesus. Scripture describes John as preparing the people of that time and place for the coming of a new reality. According to the Gospel of Matthew, John said, "One who is more powerful than I is coming after me."

What happened in Puerto Rico with Hurricane Maria signals what is coming. As the very bad San Felipe Segundo was followed by the even worse San Ciprian, there will be hurricanes after Maria. This is not a threat just to Puerto Rico, and hurricanes are not the only problem.

Nana told her grandchildren, "Listen. Listen to the wind. It is whispering of tomorrow. Watch the ants, they have stories to tell us. Seek out crazy San Juan wearing smelly skins. Ask him to tell you what will happen next. It is already happening."

There are more humans than ever before. The population of the United States has more than doubled since 1950. In 1950 about 65 percent of Americans lived in urban areas. Today more than 80 percent of Americans live in urban areas. Over the same period, a higher proportion of US population has shifted to areas with known risks for hurricane, floods, wildfire, and earthquakes.

More people more tightly concentrated in higher-risk places increases the probability of catastrophic consequences.

The story of post-Maria Puerto Rico highlights three critical relationships that amplify consequences. Complex sets of dependencies—the grid, telecommunications, fuel networks, transportation systems, and much more—enable dense populations. Failure of even one dependency can disrupt the demand-and-supply networks required to sustain human life. In the event of such failures, the proximity of dense populations to other dense sources of supply is a key component of resilience.

Puerto Rico is at least 1,300 miles by sea or air from its principal source of supply at Jacksonville. This distance—exacerbated by human-created constraints—significantly complicated recovery.

In some scenarios a shift in the San Andreas fault just east of Los Angeles will produce a thirty-foot offset between one side and the other. Bridges, roads, pipelines, and power cables will stretch and break. With luck connections running northwest of the dense urban area may persist, but thirteen million residents could suddenly be living on a heavily damaged pop-up peninsula.

The San Andreas is one of several faults running beneath the 18,000-people-per-square-mile crowded into San Francisco. That seven-by-seven-mile matrix is already a peninsula. Will connections further narrow?

Four million live along the hills and shores of the Puget Sound. The predicted large-scale Cascadia seismic event is likely to create many new islands from now-continuous urban sprawl. The single largest source of fuel—for the two million in Portland, too—is likely to be disconnected in several places.

Unlike Philadelphia, Baltimore, New York, or Boston, Los Angeles, San Francisco, Portland, and Seattle are islands of dense population separated from comparable sources of supply. So is Honolulu. What about Salt Lake City? What about your home city? What is your density? What are your crucial dependencies? How distant are you from other comparably dense sources of supply?

John was the baptizer. He invited people to wade into the middle and get soaked. In some traditions, those baptized are immersed in the river. Find the flow. Stay with the flow.

But the love will have been enough; all those impulses of love return to the love that made them. Even memory is not necessary for love. There is a land of the living and a land of the dead and the bridge is love, the only survival, the only meaning.

—Thornton Wilder, *The Bridge of San Luis Rey*

Research Bibliography

■ ■ ■

This narrative is based on several sources. In 2017 Philip J. Palin supported the FEMA response to Hurricanes Harvey, Irma, and Maria. In 2018, he conducted a series of interviews with several supply chain owners and operators who had experienced Hurricanes Harvey, Irma, and Maria. These experiences and other research were reported in two prior publications:

"Learning from H.I.M. (Harvey, Irma, Maria): Preliminary Impressions for Supply Chain Resilience." Homeland Security Affairs. September 29, 2018. https://www.hsaj.org/articles/14598.
"Supply Chain Resilience and the 2017 Hurricane Season." 2018. Institute for Public Research, CNA. October 2018. https://www.cna.org/research/hurricane-supply-chain.

DEMOGRAPHY, ECONOMY, AND WEATHER

"Ascertainment of the Estimated Excess Mortality from Hurricane Maria in Puerto Rico." 2018. Milken Institute School of Public Health. The George Washington University. August 2018. https://publichealth.gwu.edu/sites/default/files/downloads/projects/PRstudy/Acertainment of the Estimated Excess Mortality from Hurricane Maria in Puerto Rico.pdf.
"Gross Domestic Product Per Capita for Puerto Rico." 2018. FRED Economic Data. Federal Reserve Bank of St. Louis. July 20, 2018. https://fred.stlouisfed.org/series/NYGDPPCAPCDPRI.
"Major Hurricane Maria." 2018. National Weather Service. NOAA's National Weather Service. May 6, 2018. https://www.weather.gov/sju/maria2017.
"Puerto Rico: A Way Forward." 2015. Government Development Bank. June 2015. http://www.gdb.pr.gov/documents/puertoricoawayforward.pdf.

"Puerto Rico Economic Data." n.d. Economic Development Bank. http://www.bde.pr.gov/bdesite/PRED.html.

"Puerto Rico: Quick Facts." n.d. Census Bureau QuickFacts. United States Census Bureau. https://www.census.gov/quickfacts/fact/table/pr/PST045217.

"Report on the Competitiveness of Puerto Rico's Economy." 2012. Federal Reserve Bank of New York. June 29, 2012. https://www.newyorkfed.org/medialibrary/media/regional/PuertoRico/report.pdf.

"2017 Hurricane Season After-Action Report." 2018. FEMA. July 12, 2018. https://www.fema.gov/media-library-data/1531743865541-d16794d43d30 82544435e1471da07880/2017FEMAHurricaneAAR.pdf.

"2017 Hurricanes and Fires: Initial Observations on the Federal Response and Key Recovery Challenges." 2018. United States Government Accountability Office. September 2018. https://www.gao.gov/assets/700/694231.pdf.

ELECTRICAL AND CELL SYSTEMS

"Commonwealth of Puerto Rico (17–03283)." n.d. Prime Clerk. Accessed December 31, 2018. https://cases.primeclerk.com/puertorico/Home-Docket Info?DocAttribute=4417&DocAttrName=ADV.CASENO.18-00101.

Ferris, David. 2018. "HURRICANE MARIA: Puerto Rico's Grid Recovery, by the Numbers." Energywire. E&E News. February 20, 2018. https://www.eenews.net/stories/1060074219.

Fisher, Jeremy I., and Ariel I. Horowitz. 2016. "Expert Report: State of PREPA's System." Puerto Rico Energy Commission. November 23, 2016. http://energia.pr.gov/wp-content/uploads/2016/11/Expert-Report-Revenue-Requirements-Fisher-and-Horowitz-Revised-20161123.pdf.

Gallucci, Maria. 2018. "Rebuilding Puerto Rico's Power Grid: The Inside Story." IEEE Spectrum: Technology, Engineering, and Science News. IEEE Spectrum. March 12, 2018. https://spectrum.ieee.org/energy/policy/rebuilding-puerto-ricos-power-grid-the-inside-story.

"Puerto Rico Electric Power Authority: Amended & Restated Fiscal Plan." 2018. Puerto Rico Electric Power Authority. January 24, 2018. http://www.aafaf.pr.gov/assets/prepa-revisedfiscalplan-01-24-18.pdf.

"Puerto Rico's Electricity Service Is Slow to Return after Hurricane Maria." 2017. Today in Energy. US Energy Information Administration. October 24, 2017. https://www.eia.gov/todayinenergy/detail.php?id=33452.

"2017 Atlantic Hurricane Season Impact on Communications Report and Recommendations Public Safety Docket No. 17–344." 2018. Federal Communications Commission. August 2018. https://docs.fcc.gov/public/attachments/DOC-353805A1.pdf.

MARITIME LINKS AND PORTS

"Fact Sheet: Domestic Maritime Industry Dedicated to Puerto Rico." 2017. American Maritime Partnership. November 17, 2017. https://www.americanmaritimepartnership.com/press-releases/domestic-maritime-industry-dedicated-puerto-rico.

Gómez, Suárez II. 2016. "Cabotage: The Effects of an External Non-Tariff Measure on the Competitiveness of Agribusiness in Puerto Rico." Bradford Scholars. University of Bradford. 2016. https://bradscholars.brad.ac.uk/handle/10454/13464.

Grennes, Thomas. 2017. "An Economic Analysis of the Jones Act." Mercatus Center, George Mason University. 2017. https://www.mercatus.org/system/files/mercatus-grennes-jones-act-v2.pdf.

Hoffman, Jann. 2017. "Consolidation in Liner Shipping—Time Flies." Investment Country Profiles. UNCTAD. September 27, 2017. https://unctad.org/en/pages/newsdetails.aspx?OriginalVersionID=1567.

"Impact of the U.S. Jones Act on Puerto Rico." 2018. Reeve & Associates and Estudios Técnicos, Inc. July 2018. https://3snn221qaymolkgbj4a0vpey-wpengine.netdna-ssl.com/wp-content/uploads/2018/07/Report_Impact-of-the-Jones-Act-on-Puerto-Rico_FINAL2.pdf.

Klein, Aaron. 2017. "What Everyone Got Wrong about the Jones Act, Hurricane Relief, and Puerto Rico." The Brookings Institution. October 25, 2017. https://www.brookings.edu/blog/up-front/2017/10/25/what-everyone-got-wrong-about-the-jones-act-hurricane-relief-and-puerto-rico.

"Number and Size of the U.S. Flag Merchant Fleet and Its Share of the World Fleet." n.d. Bureau of Transportation Statistics. https://www.bts.gov/content/number-and-size-us-flag-merchant-fleet-and-its-share-world-fleet.

"Port Performance Freight Statistics Program: Annual Report to Congress." 2017. US Department of Transportation, Bureau of Transportation Statistics. 2017. https://www.bts.gov/sites/bts.dot.gov/files/docs/browse-statistical-products-and-data/port-performance/216906/bts-ppfsp-ar-congress-2017.pdf.

"Puerto Rico: Characteristics of the Island's Maritime Trade and Potential Effects of Modifying the Jones Act." 2013. United States Government Accountability Office. March 2013. https://www.gao.gov/assets/660/653046.pdf.

"Puerto Rico's Post-Hurricane Operations." 2018. American Shipper. March 21, 2018. https://www.americanshipper.com/news/?autonumber=70879&source=redirected-from-old-site-link.

"VIDEO: Crowley Delivers Around-the-Clock Response to Hurricane Maria in Puerto Rico." 2017. Crowley—People Who Know. October 9, 2017. http://www.crowley.com/News-and-Media/Press-Releases/VIDEO-Crowley-Delivers-Around-the-Clock-Response-to-Hurricane-Maria-in-Puerto-Rico.

FOOD, ENERGY (NON-GRID), AND WATER

Brumby, Seth, Javier Balmaceda, and Xavira Neggers Crescioni. 2018. "How Burger King Fed Storm-Ravaged Puerto Rico, and Made a Killing." Debtwire Investigations. May 18, 2018. http://investigations.debtwire.com/how-burger-king-fed-storm-ravaged-puerto-rico-and-made-a-killing.

Burgos Alvarado, Cindy. 2017. "Puerto Rico Consumer Affairs Dept: 500 Gas Stations Operating." Caribbean Business. September 27, 2017. https://caribbeanbusiness.com/puerto-rico-consumer-affairs-dept-500-gas-stations-now-open.

"Company Feels Responsible for Feeding Puerto Rico." *El Nuevo Dia.* October 2017. https://www.elnuevodia.com/brandshare/b-fernandez/nota/empresaboricuasesienteresponsabledealimentarapuertorico-2371158/

Cunningham, Avery. 2017. "National Guard Purifies and Distributes Water." Vermont National Guard. October 26, 2017. http://vt.public.ng.mil/News/News-Article-View/Article/1416945/national-guard-purifies-and-distributes-water.

Dorell, Oren. 2017. "With Long Lines for Food, Water and Fuel and No Electricity, Puerto Ricans Help Each Other." *USA Today.* October 1, 2017. https://www.usatoday.com/story/news/nation/2017/10/01/puerto-rico-want-and-generosity/720663001.

Fausset, Richard, Frances Robles, and Deborah Acosta. 2017. "Minus Electrical Grid, Puerto Rico Becomes Generator Island." *New York Times.* October 7, 2017. https://www.nytimes.com/2017/10/07/us/puerto-rico-power-generators.html.

"Government Agencies Not Taking Action against Dangers of Electric Generators in Puerto Rico." 2017. Latino USA. Center for Investigative Journalism. November 16, 2017. https://latinousa.org/2017/11/13/government-agencies-not-taking-action-dangers-electric-generators-puerto-rico.

"Hurricanes Maria, Irma, and Harvey: September 21 Morning Event Summary (Report #40)." 2017. Infrastructure Security & Energy Restoration. US Department of Energy. September 21, 2017. https://www.energy.gov/sites/prod/files/2017/09/f36/Hurricanes Maria Irma & Harvey Event Summary morning September 21, 2017.pdf.

"Hurricanes Nate, Maria, Irma, and Harvey Situation Reports." n.d. Department of Energy. https://www.energy.gov/ceser/downloads/hurricanes-nate-maria-irma-and-harvey-situation-reports.

Malik, Naureen S, and Catherine Traywick. 2017. "Tiny Fleet Converges on Puerto Rico With Lifeblood Diesel." Bloomberg. September 27, 2017. https://www.bloomberg.com/news/articles/2017-09-27/tiny-fleet-converges-on-puerto-rico-to-deliver-lifeblood-diesel.

"National Water Information System: Web Interface." n.d. US Department of the Interior. United States Geological Survey. https://nwis.waterdata.usgs.gov/pr/nwis/uv/?cb_00060=on&cb_00065=on&format=gif_default&site_no=50044810&period=&begin_date=2017-09-19&end_date=2017-10-03.

Parraga, Marianna. 2017. "Buckeye Resumes Operation at Yabucoa, Puerto Rico; Tankers. . . " Reuters. October 3, 2017. https://www.reuters.com/article/us-usa-puertorico-buckeye-partners/buckeye-resumes-operation-at-yabucoa-puerto-rico-tankers-discharge-fuel-idUSKCN1C8273.

"Puerto Rico: Territory Profile and Energy Estimates." n.d. US Energy Information Administration. https://www.eia.gov/state/data.php?sid=RQ#Supply.

Respaut, Robin, Dave Graham, and Jessica Resnick-Ault. 2017. "For Desperate Puerto Ricans, Fuel a Precious Commodity." Reuters. September 27, 2017. https://www.reuters.com/article/us-usa-puertorico-fuel/for-desperate-puerto-ricans-fuel-a-precious-commodity-idUSKCN1C216B.

"Revised Fiscal Plan to Incorporate Modifications to the Certified Fiscal Plan as a Result of the Impact of Hurricanes Irma and Maria." 2018. Puerto Rico Aqueduct and Sewer Authority, Government of Puerto Rico. March 23, 2018. http://www.aafaf.pr.gov/assets/prasa-revised-fiscal-plan-march-2018—draft.pdf.

Smith, Jennifer, Paul Page, and Arian Campo-Flores. 2017. "Puerto Rico Aid Trickles In." *Wall Street Journal.* September 28, 2017. https://www.wsj.com/articles/food-fuel-and-medical-supplies-slowly-distributed-across-puerto-rico-1506622283.

"Surface-Water, Water-Quality, and the Ground-Water Assessment of the Municipio of Comerío, Puerto Rico, 1997–99." 2001. US Department of the Interior, US Geological Survey. 2001. https://pubs.usgs.gov/wri/wri01-4083/pdf/wri014083.pdf.

Williams, Susan E., Stacy Cagle Davis, and Robert Gary Boundy. 2018. "Transportation Energy Data Book: Edition 36." Energy and Transportation Science Division, August. https://doi.org/10.2172/1410917.

NETWORK SCIENCE

Kirschner, Marc, and John Gerhart. 1998. "Evolvability." *Proceedings of the National Academy of Sciences of the United States of America*, July, 8420–27. https://doi.org/10.1073/pnas.95.15.8420.

Lewis, Ted G. 2009. *Network Science: Theory and Practice.* Hoboken, NJ: Wiley.

Lewis, Ted G. 2013. "Cognitive Stigmergy: A Study of Emergence in Small-Group Social Networks." *Cognitive Systems Research* 21 (March): 7–21. https://doi.org/10.1016/j.cogsys.2012.06.002.

Sabrin, Kaeser M., and Constantine Dovrolis. 2018. "The Hourglass Effect in Hierarchical Dependency Networks." *Georgia Institute of Technology 5* (04): 490–528. https://arxiv.org/pdf/1605.05025.pdf.

Ulusan, Aybike, and Ozlem Ergun. 2018. "Restoration of Services in Disrupted Infrastructure Systems: A Network Science Approach." *PLOS One* 13 (2). https://doi.org/10.1371/journal.pone.0192272.

Valckenaers, Paul, Hendrik Van Brussel, Martin Kollingbaum, and Olaf Bochmann. 2001. "Multi-Agent Coordination and Control Using Stigmergy Applied to Manufacturing Control." Multi-Agent Systems and Applications Lecture Notes in Computer Science, 317–34. https://doi.org/10.1007/3-540-47745-4_15.

Oh, I do not understand. It is just circumstance. I must be what I must. Do not try to understand either. Don't think about me. . . . Just forgive, that's all. Just try to forgive.
—Thornton Wilder, *The Bridge of San Luis Rey*

Che si può fare? What Can I Do?

■ ■ ■

Verse by Gaudenzio Brunacci, set to music by Barbara Strozzi, whistled by Manuel, see page 49.

What can I do?
If stars without pity, work against me;
If heaven offers no trade
Of peace for my pain,
What can I do?

What can I say?
If stars rain disasters upon me;
If love does not grant a moments breath,
to relieve all my suffering,
What can I say?
(original Italian text)

Che si può fare?
Le stelle ribelle non hanno pietà;
se 'l cielo non dà un influsso
di pace al mio penare,
che si può fare?
Che si può dire?
Dagl'astri disastri mi piovano ognor;
se perfido amor un respiro diniega
al mio martire,
che si può dire?

Source: https://youtu.be/aDBPfhG-gVk

Index

∎ ▦ ▦

About the Author

■ ■ ■

PHILIP J. PALIN is the son and grandson of grocers. Following a career in entrepreneurship and education, since 2008 Mr. Palin has researched, written, and consulted on supply chain resilience, catastrophe preparedness, and private-public relationships. Mr. Palin serves as the principal investigator for supply chain resilience with the Institute for Public Research at CNA Corporation and staff consultant on supply chain resilience for the Resilient America Roundtable (National Academies of Sciences, Engineering, and Medicine), and is a subject matter expert for the FEMA National Integration Center Technical Assistance Program in Supply Chain Resilience.

www.ingramcontent.com/pod-product-compliance
Lightning Source LLC
Chambersburg PA
CBHW051252050326
40689CB00007B/1162

* 9 781538 118207 *